Yale Agrarian Studies Series

James C. Scott, Series Editor

The Agrarian Studies Series at Yale University Press seeks to publish outstanding and original interdisciplinary work on agriculture and rural society—for any period, in any location. Works of daring that question existing paradigms and fill abstract categories with the lived experience of rural people are especially encouraged.

—James C. Scott, Series Editor

James C. Scott, *Seeing Like a State: How Certain Schemes to Improve the Human Condition Have Failed*

Steve Striffler, *Chicken: The Dangerous Transformation of America's Favorite Food*

James C. Scott, *The Art of Not Being Governed: An Anarchist History of Upland Southeast Asia*

Timothy Pachirat, *Every Twelve Seconds: Industrialized Slaughter and the Politics of Sight*

James C. Scott, *Against the Grain: A Deep History of the Earliest States*

Jamie Kreiner, *Legions of Pigs in the Early Medieval West*

Christian Lund, *Nine-Tenths of the Law: Enduring Dispossession in Indonesia*

Shaila Seshia Galvin, *Becoming Organic: Nature and Agriculture in the Indian Himalaya*

Michael R. Dove, *Bitter Shade: The Ecological Challenge of Human Consciousness*

Japhy Wilson, *Reality of Dreams: Post-Neoliberal Utopias in the Ecuadorian Amazon*

Aniket Aga, *Genetically Modified Democracy: Transgenic Crops in Contemporary India*

Ruth Mostern, *The Yellow River: A Natural and Unnatural History*

Brian Lander, *The King's Harvest: A Political Ecology of China from the First Farmers to the First Empire*

Jo Guldi, *The Long Land War: The Global Struggle for Occupancy Rights*

Andrew S. Mathews, *Trees Are Shape Shifters: How Cultivation, Climate Change, and Disaster Create Landscapes*

Francesca Bray, Barbara Hahn, John Bosco Lourdusamy, and Tiago Saraiva, *Moving Crops and the Scales of History*

Deborah Valenze, *The Invention of Scarcity: Malthus and the Margins of History*

Brooks Lamb, *Love for the Land: Lessons from Farmers Who Persist in Place*

Michael R. Dove, *Hearsay Is Not Excluded: A History of Natural History*

For a complete list of titles in the Yale Agrarian Studies Series, visit https://yalebooks.yale.edu/search-results/?series=yle10-yale-agrarian-studies-series

Nature, Culture, and Race in Colonial Cuba

Lee Sessions

Yale
UNIVERSITY PRESS
NEW HAVEN AND LONDON

Publication of this book has been aided by a grant from the Millard Meiss Publication Fund of CAA, and also with assistance from the foundation established in memory of Amasa Stone Mather of the Class of 1907, Yale College.

Set in Fournier type by Integrated Publishing Solutions.
Printed in the United States of America.

ISBN 978-0-300-26794-5 (hardcover : alk. paper)
Library of Congress Control Number: 2023944737
A catalogue record for this book is available from the British Library.

This paper meets the requirements of ANSI/NISO Z39.48-1992 (Permanence of Paper).

10 9 8 7 6 5 4 3 2 1

Contents

Acknowledgments

I am deeply grateful to so many people in so many different communities for their support, perhaps most of all the scholars from all walks of life who came before me in the archives, the museum storerooms, and the streets and cafes of Havana, keeping these histories alive. I have always been forthright about the fact that the stories that I write about are not my own. I have no ancestral ties to Cuba or to the Caribbean that I know of. I am only connected to these stories in the ways I have benefited from white privilege throughout my life and, by doing so, have contributed to the reification of racial division and inequity today. I hope that by studying this process in a different time and place, I can help show how we might tear down these divisions, and I pay all respects possible to all activists that have been doing this crucial work of dismantling white supremacy for many centuries.

I learned to see in art museums and in galleries. Joe Ketner and the staff of the Rose Art Museum at Brandeis University introduced me to the many layers that could be contained in a single work of art. Ro Lohin from Lohin Geduld Gallery has taught me so much about painting and about life, and she remains an inspiration. At the Brooklyn Museum, I found my love of art and my love of the people this art could serve expanding through the generous mentorship of Richard Aste and Teresa A. Carbone. Pat Dignan, Susan Handwerker, and Alicia Degener at the Brooklyn Waterfront Artists Coalition showed me the power of a true art community and the beautiful objects that such a community makes possible.

The incredible scholars of the LuEsther T. Mertz Library and the Humanities Institute of the New York Botanical Garden, including Rashad Bell, Samantha D'Acunto, and Stephen Sinon, truly made this entire book possible. I am particularly grateful for the friendship, generosity, and enthusiasm of Vanessa Sellers at the Humanities Institute of the New York Botanical Garden, whose comments made every part of this book more rigorous and richer. And, most recently, Rodrigo Moura, Susanna Temkin, Daniel Silva, Michelle McVicker, and Alexia Arrizurieta have welcomed me into la familia of El Museo del Barrio and reminded me of the excitement, intellectual possibility, and pure joy of working closely with a wonderful collection of objects and connecting with a fantastic neighborhood.

My undergraduate adviser Dave Jacobson of Brandeis University was one of the first professors to walk me through an ambitious project and teach me the joys of close listening and careful argumentation. I was then blown away by the complexity, rigor, and generosity of the faculty at the Institute of Fine Arts during my M.A. studies. Thomas Crow, Finbarr Barry Flood, Alexander Nagel, and Robert Slifkin, among many other brilliant scholars there, truly opened my eyes to the possibilities of art history. Most of all, Edward Sullivan has been a friend and mentor for many years, from his first encouragement to study Latin American art to his brilliant responses to these chapters. Ned Cooke of Yale University enriched this project at every stage, and it would not have been possible without his insightful suggestions, his clear vision, and his unwavering support. I am also extremely grateful for the feedback I received on parts of this project at various stages from other members of the Yale University History of Art faculty, especially Carol Armstrong, Tim Barringer, Marisa Bass, Jacqueline Jung, Jennifer Raab, and Nicola Suthor.

I am more indebted than I can say to scholars from other institutions for their advice, expertise, and personal support. I especially thank Alejandro Barro of Universidad de la Habana, Joseph R. Hartman of the University of Missouri–Kansas City, Ray Hernández-Durán of the University of New Mexico, Luis Gordo Pelaez of California State University, Fresno, Agnes Lugo-Ortiz of the University of Chicago, Reinaldo Funes Monzote of Universidad de La Habana, Paul Niell of Florida State University, and Lilian Garcia Roig of Florida State University. Erica Moiah James of the University of Miami, in particular, has given me wise advice and steadfast encouragement for many years, and I am incredibly grateful.

My editor, Jean Thomson Black at Yale University Press, led me through this project with good cheer, gentle pushes, and boundless intellectual curiosity. I am so appreciative of your vision and your support. I am eternally grateful to the readers who gave insightful feedback on this manuscript and gave so much shape and rigor to the final book.

I was able to conduct preliminary research for this project with the generous support of the Council on Latin American and Iberian Studies at Yale

University and funding from the Yale History of Art Department, and I wrote an early draft of my third chapter during a residency at the Wertheim Study at the New York Public Library. The MacMillan International Dissertation Research Fellowship enabled me to complete the bulk of my research, and I am deeply grateful for their support. I completed the majority of the revisions to this manuscript while serving as an Andrew W. Mellon Postdoctoral Fellow at the Humanities Institute of the New York Botanical Garden.

I presented parts of this book and received invaluable feedback at the following panels and conferences: "Liquidities: Seascapes as Subject and Method" panel at the College Art Association Annual Conference, February 2023; "Los Ingenios and the End of Cuban Sugar" panel at the College Art Association Annual Conference, February 2021; "Space, Science, and Slavery in Havana's Botanical Garden" panel at the 4th Annual International Conference of the Society of Architectural Historians, February 2021; "Havana: New Research and Critical Reflections on an Urban Palimpsest through Five Centuries" panel at the College Art Association Annual Conference, February 2020; the 2019 Pre-Latin American Studies Association Workshop; the Graduate Students' Research on Cuba conference at Yale University, the Emerging Scholar Symposium of Latin American Art, and the Yale American Art History Graduate Student Symposium in 2018; "Entre la ciencia y el arte: La naturaleza cubana en La Sagra, Humboldt y Akerman," at the Antonio Núñez Jiménez Foundation in Havana in 2017; and the Graduate Student Conference on Latin American and Caribbean Studies at Yale University in 2016.

I am extremely grateful to the librarians, administrators, and other staff at the following institutions, who always made me feel welcome and who deepened my research immeasurably: the Museo de Historia Natural Felipe Poey, the Instituto de Literatura y Lingüística, the Instituto de Historia de Cuba, the Archivo Nacional de Cuba, and the Biblioteca Nacional José Martí in Havana; the Real Jardín Botánico, the Museo Nacional de Ciencias Naturales, the Museo Naval, and the Archivo Histórico Nacional in Madrid; the Archivos General de las Indias in Seville; the New-York Historical Society and the New York Public Library in New York; and the Beinecke Library in New Haven.

Because they are fighting every day against the precarity of academia and because their support and strength carried me through this process, I am also profoundly grateful to my fellow members of UNITE HERE–Local 33.

My family, those I was born with and those I chose and that chose me, inspires and sustains me. I've talked the ears off of Masha Bezlepkina, Ami Imai

Brett, Miki Foster, Ben Goodman, Yeshim Iqbal, Jana Littleton, Duja Michael, Barbara Sabitzer, Matt Sorensen, Maura VanVoris, and Brendan VanVoris about this book for years and each of their voices can be found on every page of it. Reginald and Nai-Chia Luke have welcomed me, cheered for me, grounded me, and created a second home for me. Joey Nicholson listened to me, made me laugh, and brought me down to earth. Antonia Pocock nurtured me and inspired me with her boundless love of art. Kate Schwartz knew what I was capable of before I knew it myself and taught me to live a life of adventures. Ryah Schwartz reminded me to live in the moment and to stay in the water for one more wave. Alice Sessions inspired me with her ferocious intellect and constant learning and always encouraged me to stay true to myself and to fight for what I believe in. Mike Sessions taught me to dive into what excites my curiosity and to constantly orient myself toward justice and toward love. Roger Sessions showed me the connections between all the parts of our universe and showed me how to build communities of love and joy.

And of course, none of this would have been possible without Bing Luke. He was at my side for every word and celebrated every milestone with me, made me laugh and made me take walks and made coffee every morning. I'm so grateful to be drinking that coffee with you.

Nature, Culture, and Race in Colonial Cuba

Introduction

An article published in Havana's leading scientific magazine in 1836 brought the reader's attention to an issue that was paramount in the minds of many in the city at the time. Increasing the white population of Cuba was a matter, the unnamed author wrote, "that demands our total attention and care," "one of the most urgent necessities of this nearly deserted island," and a topic that "merits the attention of all those who feel a patriotic love in their breast." The author went on to describe Cuba's natural resources, including the fertility of the soil that allowed for the production of tobacco, beeswax, honey, a variety of woods, and high-quality sugar and coffee. The article also described Cuba's advantageous geographical position for exporting these products. These advantages together presented "a rich and florid area of promise" that could help attract white immigrants to the island to increase the aforementioned white population.[1]

This article outlining the fecundity and geographical advantages of the island was followed in the same magazine by a discussion of the importance of increasing agricultural production across Cuba and some of the technological advances that would facilitate this increase. "If there is no wealth more secure than that wealth with which nature rewards the labor and efforts of man, then why is the science of the honest and profitable operation [of agriculture in Cuba] almost entirely unknown?," the also unnamed author asked.[2]

Why this sudden intense interest in white population, agriculture, and the links between the two? One hint is found in another series of articles from the same year asking the question, "can we make up for the lack of slave labor in a sugar plantation?" The (again, unnamed) author of these articles discussed increasing challenges to slavery in Cuba, including the end of the legal slave trade through treaty with Britain in 1820, rising impediments to the illegal slave trade that supplied the expanding plantations of Cuba with enforced labor, and the spread of emancipation across the Caribbean region. The article concluded, perhaps unsurprisingly, that the solution to the impending emancipation of enslaved people in Cuba should be to increase white immigration to Cuba. "Since our principal object is to increase the white population, it seems quite convenient to put most of our efforts of recruitment into married couples of fertile age," its author wrote.[3]

This sample of articles, just a few of the hundreds published on these topics during the 1830s and 1840s in Cuba, speaks to the intense racial anxiety of white elites in Havana at the time. In addition to simple racial animus, this anxiety was driven by political uncertainty and expanded Spanish imperial force on the island, the rapid increase of the sugar industry and expansion of legal and illegal enslavement to supply this industry with labor, and the political and economic gains made by free black and brown Cubans. White Cubans imagined themselves to be threatened on all sides—by their government, by the changing demographics of their island, and by shifting social norms around race. Their anxiety, in turn, drove their near obsession with white immigration to "balance the demographics" of Cuba. This immigration, recall, was considered by some to be "one of the most urgent necessities of this nearly deserted island."

What this set of articles also shows is that this white immigration was tied in the minds of these white elites to Cuban nature and science. Publicizing Cuba's natural resources to audiences in Europe and (to a lesser extent) the United States would, it was believed, entice white settlers to the island, improved Cuban agriculture would feed and occupy them, and the trade networks that Cuba's location had access to would distribute the products that they grew. As one author wrote as early as 1817, "The island of Cuba, through its geographic situation, appears *destined by nature* to be the universal deposit of all the commercial nations; and for the products of the soil to increase the delights of *the civilized people* [of the world]."[4] According to this model, then, one of the primary goals of Cuban naturalists and scientists was to display Cuban fecundity to entice white settlers to the island's shores.

The first two-thirds of the nineteenth century saw a flourishing of various projects and spaces that were intended for this display. These included a new natural history museum and botanical garden in the city of Havana as well as many ambitious illustrated publications showcasing Cuban botany. These also included agricultural exhibitions, lithographs celebrating Cuban landscapes, and public and semi-public gardens around the city. Together, these projects represented the island of Cuba as a lush paradise just waiting to enrich white migrants. In this book I examine each of these projects in detail, discussing them in the context of similar projects both in Cuba and across the world at the time.

In a broader sense, then, this book examines how the natural world, material culture, and political power are entangled. I use the case study of Cuba to examine how elites in a particular society wielded power through built spaces and printed volumes. I also use the case study of Cuba to examine how these elites constructed and reconstructed the imagined communities that they belonged to, including their own racial identities. The natural world was central to both of these processes. Elites in Havana used representations of local nature to signal identity to each other, to reify their power, and to create futures in which this power was expanded. The material culture of Cuban nature—wooden cases holding museum specimens, prints depicting Cuban flora, fauna, and landscapes, the built environment of the botanical garden—was essential to white elite power on the island in the decades before the wars of independence. By examining this case study in detail, we can understand how nature has been wielded to consolidate power at many other moments in history and at many other locations.

At 42,426 square miles, the main island of Cuba is the largest landmass in the Caribbean Sea and is about one hundred miles from Florida at its closest point. Cuba remained a colony of Spain throughout the period that I examine. However, over the course of the nineteenth century, Cubans' experience of Spanish rule shifted significantly, with some periods of relatively little crown interference interspersed with moments of intense restriction on political rights. The century also witnessed a number of independence movements of varying forms and popularities. These independence movements often stretched across racial and geographical boundaries, as Dalia Antonia Muller, Louis A. Pérez Jr., and others have discussed.[5] In this book, I focus on the period that began with the rise of the Cuban sugar economy in the early 1820s and I end with the decline of this industry and the wars of independence of the late 1860s. I chose this time period because, as I will discuss at length, it was marked by massive social, ecological, and political changes in Cuba and because these changes were often put into motion, dramatized, and thought through in the expanding field of Cuban academic science.

Dubbed "Key to the New World and Rampart of the West Indies" by royal decree in 1634, Havana was the bustling center of Cuban life. By the middle of the eighteenth century, Havana was the third largest city in the Americas (after Lima and Mexico City) and was a key port for the Spanish empire, as all *flota de indias* ships departing Lima, Veracruz, or Cartagena stopped there to rendezvous into convoys to sail together to Spain. Havana's ports only increased in importance as the city's wealth grew and as shipping intensified during the sugar boom of the nineteenth century. Alexander von Humboldt described "the city half-hidden behind a forest of masts and sails" upon entering the harbor in 1800.[6] The inhabitants of the city maintained social and economic ties not just to Spain but to sites throughout the Caribbean basin and the Atlantic world. These ties were made up of and maintained by the movement of objects as well as people, and these objects included books, paintings, drawings, and prints as well as furniture, clothing, and innumerable other items. Inhabitants of Havana, then, used these objects, this material culture, to demonstrate loyalties and aspirations and broadly to negotiate their identities within these networks.[7]

The white elite intellectuals that I examine understood the population of Cuba to be composed of white people, free people of color, and slaves. These are the terms that the Spanish-born scientist Ramón de la Sagra y Peris, for instance, used on a map that he created in 1843.[8] Sagra estimated that Cuba at the time was 45 percent white, 15 percent free people of color, and 40 percent enslaved people of African descent—estimates that historians today would generally concur with. These simple percentages, however, belie a more nuanced story. The definitions of each of these groups shifted significantly from year to year, and one's membership in them was even subject to negotiation. For example, someone of mixed race could have applied for a *gracias al sacar,* a royally granted "certificate of whiteness," which was fairly common in the early nineteenth century. This certificate allowed the recipient to participate in social structures such as universities, church ordination, or medical practice that were limited to white Cubans.[9]

Furthermore, race and power intersected closely with changes in imperial rule. David Sartorius, for instance, discusses how free and enslaved people of African descent in Cuba claimed loyalty to Spain at different points in the nineteenth century to secure additional social mobility and legal power.[10] Rebecca J. Scott, Bonnie A. Lucero, and Adriana Chira have also written about how Cubans of African descent negotiated political status inside and outside of the intersecting legal systems of nineteenth-century Cuba.[11] Recent scholarship that focuses on the unrest of the 1840s in Cuba continues to nuance these discussions of power, violence, and race in the island's history.[12]

Narratives of race in Cuba are further complicated by different accounts of the fate of the island's indigenous population. Before the arrival of Columbus, Cuba was inhabited by two primary groups, now known as the Taíno and the Ciboney, who maintained complex social, political, and economic systems within the island and across the region. For many centuries, these groups were discussed as if they had been entirely eradicated by violence, enslavement, and especially disease in the years after the European invasion. However, this narrative of genocide has been firmly rejected in recent years both by geneticists who have found significant DNA evidence for continued indigenous presence in the region and by individuals and communities themselves that identify as Taíno.[13] Scientific knowledge held by Taíno communities has shaped Cuban foodways, medicine, and botany for centuries, even when their continued presence was not acknowledged, and continues to drive medical innovation today.[14] There are no records in the archives of the scientists of elite white Havana being in contact with Taíno knowledge-holders, but it is almost certain that indigenous science played some role in the account that follows, difficult as that role might be to trace.

The category of "white" was also complicated. Throughout the period I examine, criollos (people of European descent who were born in Cuba) and peninsulares (people of European descent born in Spain) were both considered white. Sherry Johnson describes how in the late eighteenth century, the Spanish monarchy manipulated systems of official privilege and intermarriage laws between criollos and peninsulares and, by doing so, "implicitly sent the message that Cuba's wealthiest criollo families were worthy—perhaps favored—citizens of the Spanish empire."[15] Many of the owners of sugar plantations, in particular, were criollo but had access to extreme wealth as well as the titles this wealth could purchase. In practice, then, even in moments of debates over the rights of citizenship, these individuals often had significantly more power than even highly placed peninsular government officials. As Johnson and others have noted, however, this privileged status of criollos came under threat at various moments throughout the nineteenth century, especially during the constitutional reforms of the 1820s. During times of increased imperial control, peninsulares were often granted more political rights than criollos, and the frustration of many criollos at this situation drove many of the independence movements of the century.

Furthermore, immigrants of various ethnic backgrounds from around the world, including other Caribbean islands, the United States, the Canary Islands, the Yucatan, and Asia, had "whiteness" bestowed upon them or taken away. Consuelo Naranjo Orovio and Armando García González discuss this process at length and note how science was used to reinscribe or shift racial status.[16] I take up this examination of how whiteness was defined differently by various groups at different points in Havana's history and how this shifting definition intersected with other developments in culture.

Across these shifting sands of racial definitions, mobility, and negotiation, one constant structured, in some way, the life of every person in Cuba in the nineteenth century—sugar. The role of sugar in the economic and political life of the island during this period cannot be overstated. Cuba contained only a smattering of sugar plantations when the Haitian Revolution began in August 1791. By September of the same year, the revolution had destroyed hundreds of plantations across that neighboring island. Despite their small numbers at the time, Cuban plantation owners recognized new opportunities in the sugar market and were determined to take full advantage of the turmoil. They advocated vociferously for government support of the still nascent Cuban sugar industry, and they were successful. The number of sugar plantations in Cuba nearly doubled from 529 in 1792 to 1,000 in 1827, and sugar production increased from 19,000 tons in 1792 to 73,000 tons in 1829, and then again to 144,000 tons in 1846.[17]

In fact, every scientific project in this book was linked in some way to sugar interests. Most of them were funded almost entirely by sugar profits. And, of course, many Cuban scientists dedicated their lives to increasing the yield of sugarcane plants and to improving the methods by which these plants were processed. The model farms of the botanical garden of Havana, for instance, included a section of sugarcane to show farmers visiting the garden, including any potential immigrants, the latest technological developments in growing sugar. However, much of Cuban science also pushed back in various ways against the hegemony of sugar. Many of the individuals that I discuss expressed grave concern about the dependence of the Cuban economy on a single crop. Many others drew attention to the massive deforestation and other forms of ecological destruction that followed the spread of sugar plantations across the island. And, of course, it was readily understood that sugar was one of the main drivers of the continuing increase of the black population of Cuba, which in turn was a primary cause of concern for many white Cuban elites. Many of the scientific projects that I discuss were meant to present a possible Cuba apart from the sugar Cuba that had come to the fore. Artists and intellectuals used the visual culture of science to imagine alternative ecologies and social structures at this time of extreme ecological destruction and social unrest.

Although the sugar plantation system made up the vast majority of economic activity and intellectual focus at the time of my study, several other important industrialized agriculture systems expanded during the nineteenth century, both on Cuba and across the region. Fernando Ortiz was one of the first scholars to bring the Cuban sugar and tobacco industries into conversation, for example, and used the latter to illustrate still-central theories of transculturation and knowledge dissemination through the meeting of cultures.[18] Much has also been written recently on Cuban coffee production and how the coffee plantation reinscribed the power systems of the sugar plantation but also created new modes of knowledge production, agricultural experimentation, and the violence of enslavement.[19] Finally, sugar production in Cuba was, of course, closely linked to wider developments in colonial plantocracies worldwide.[20] I mention some of these links in the chapters that follow, as when, for example, French engineers came to Cuba in the 1840s to sell new types of equipment to plantation owners on the island. These wider fabrics of plantation agriculture, alternative cash crops, and global ties all provide important context for the chapters that follow.

Perhaps of equal importance to these global contexts for the makers of the scientific projects that I examine was the exclusion of Cubans of color from the agricultural and intellectual future of the island. I assume throughout this book that much of the production of science, even if we use its popular definition as knowledge about the physical world produced by testing hypotheses, written up, and circulated, was driven by the knowledge and expertise held by Cubans of color, both enslaved and free.[21] This was true from the earliest European expeditions to Cuba, which depended on enslaved guides to move through space and to gather the data they desired, to the botanical garden of Havana, planted and maintained almost entirely by enslaved workers, to the written accounts of Cuban botany that synthesized generations of local knowledge. The archives that I examine in this book, however, only very rarely mention an "esclavo," "negro," "marron," or other racialized term. What I found instead in these archives and texts was a strong impulse to continually reify whiteness. Coded language was used constantly by white Cuban scientists and intellectuals to separate white from black, personhood from non-personhood. "Rural workers," "citizens," "colonists," as well as many other terms, would have been understood at the time to mean only white people. References to these rural workers became almost an invocation to protect the wealth and position of white elite Habaneros from the imagined ever-present danger of slave uprising. Indeed, these scientific projects, illustrations, and spaces served as one of the primary ways that white elite Cubans excluded brown and black people from political possibility. This denial was fueled by many fears and concerns—white elite fear of slave uprising and a recurrence of the events on neighboring Haiti, plantocracy's anxiety about dependence on black labor and the inevitability of emancipation, and perhaps even white scientists' complicated feelings about their own dependence on black knowledge and black labor.[22]

So, for instance, when a botanical garden presented a model family farm with small-scale cash crops and suggested that this farm would be run by white immigrants who were being wooed with unlimited land rights for just such a project, the accompanying implication was that these farms should not be made available to emancipated black workers when slavery in Cuba ended. When natural history books told tales of rural white workers collecting specimens and contributing knowledge to the publication, the implication was that black workers, by contrast, held no knowledge of scientific interest and were incapable of participating in the scientific project. When lithographs showed shimmering fields of sugar stretching over the Cuban landscape with enslaved black workers rendered only as pieces of machinery in the plantation's mechanism, the implication was that these peoples would cease to exist without this plantation system, fading away into the green hills beyond.

The question of how science, race, and the environment have co-defined one another at foundational moments of our modern world has been of particular interest recently in the ecological humanities. Scholars including S. J. Watts and Richard Sheridan have examined how fields such as climate determinism, natural history, and tropical medicine worked to define and structure blackness in the eighteenth and nineteenth centuries.[23] Much of the language that has been used by Europeans over the centuries of colonization derived from the concepts of danger, fecundity, excess, and the problematic entanglement of the human body and the natural world that scientists used when trying to understand how to protect the European imperial body in new tropical climates.[24] Other scholars have discussed how ecological thought and scientific metaphor made possible expansive black experiences and black futures across the Atlantic in the long nineteenth century. As Ikuko Asaka notes, for instance, black theorists in Canada and elsewhere reframed European models of climate determinism to create "alternative histories of racial science," radical geographies that claimed black political and moral rights to expanded land ownership and self-determination in temperate climates.[25] What these studies leave out, however, is what role science and ecology also played in defining whiteness during periods of intense social upheaval.

Through an art historical examination of the material culture of science in nineteenth-century Cuba, I also seek to address a related lacuna in existing Cuban historical scholarship. Scholars have, in recent years, done excellent work examining how black and brown Cubans navigated the rapidly shifting power flows of the nineteenth century, as discussed above.[26] These nuanced examinations, however, often leave out the interrelated construction of Cuban whiteness that accompanied these shifts. Scholars of race all agree that whiteness, just like blackness, was a constructed concept that had to be defined and redefined through a wide range of means.[27] The mechanisms by which whiteness was defined at a particular time and in a particular place remain understudied, however. How historical discourses of natural history and botany have helped to define whiteness during these moments, furthermore, is a potentially rich source of information to mine.[28] Consuelo Naranjo Orovio and Armando García González's *Racismo e Inmigración en Cuba* as well as García González's later *El Estigma del Color* are notable exceptions to this lacuna, with a nuanced and detailed examination of how scientific discourses were used in the nineteenth century to discuss race in Cuba, to encourage white immigration, and, in the end, to reinforce white supremacy in different ways throughout this period.[29] I expand their examination to the spaces and objects that worked in these modes.

One of the primary starting points for my work is Sibylle Fischer's book *Modernity Disavowed.*[30] Fischer argues that white Cuban society in the first half of the nineteenth century was largely defined by a constant fear that the events of the Haitian Revolution would recur on the island of Cuba. She notes that this racial anxiety was so intense that the Haitian Revolution, and the possibility of uprising in general, became actually unspeakable. I seek to show how this unspeakable terror drove what we now think of as neutral scientific production, including agricultural innovation and experimental botany. Across Cuba, as the sugar plantation system metastasized, as dozens of collectors discovered new species all the time, and as white landowners clear-cut ancient forests to feed the sugar boilers, the threat of uprising remained in the background of every print and every museum case.

In the field of art history, on the other hand, scholars have increasingly turned their attention to how art shapes and embodies power, agency, knowledge, and identity. In this social turn, pressing questions have arisen about the relationship between the humanities and the natural world. Namely, how does material culture make certain epistemologies of nature and the environment possible? Lorraine Daston, Peter Galison, Horst Bredekamp, and Ludmilla Jordanova, among others, have explored how various visual techniques have been used in scientific illustrations over the centuries to confer particular truth claims.[31] Additionally, recent studies of the discourses around vision technologies such as microscopy shed light on wider debates over ambiguity, the possibility of knowing, and the relationship between scientific knowledge and power.[32]

The case study of Cuba in the nineteenth century reveals how the objects and built spaces created by natural historians in fact drove new popular understandings of local nature itself. By doing so, the visual culture of natural history had real political effects. These images of nature were not neutral but became sites of exclusion and privilege. I come to this material from a background in art history, and the close formal analysis of that discipline forms the core of my study. Art history contends that objects and spaces shape the world around them, and furthermore that what these objects and spaces look like and how they are structured affect how they do this work. The visual structure of a print showing a specific plant species, for instance, affects how contemporary people at the time understood the information in that print. This visual structure of a single print reflected and shaped, in a small way, how people at the time understood the world around them as a whole.

To draw attention to this process, throughout this book I give detailed descriptions of various prints, drawings, and plans, and I try to recreate the spaces of natural history that also formed part of the material culture of science in Havana in the mid-nineteenth century. These objects and these spaces, precisely because of what they looked like and how they were structured, shaped the political possibilities of Cuba at this important moment. In discussing objects (including prints and drawings) that exist today, I use the art historical present tense to describe their qualities (for example, "The entrance to the garden is at the bottom of the page and a wide central path leads away from this entrance"). This convention may be less familiar to readers who do not come from an art historical background.

In addition to engaging in this visual analysis, throughout the book I put these objects and spaces in conversation with texts of the period. I found a rich source of material in the monthly publications put out by the Sociedad

Patriótica, for instance. Known at various times as the Sociedad Patriótica de la Habana, Real Sociedad Económica de la Habana, Reales Junta de Fomento y Sociedad Económica de la Habana, and Real Sociedad Económica de Amigos del País de la Habana, this group remained one of the most important characters in the story of Cuba's nineteenth century. To avoid confusion, I refer to the group as the Sociedad Patriótica or simply the Sociedad. The Sociedad ranged in size over the course of the century from dozens to hundreds of members and was composed of sugar barons, political officials, middle-class professionals, and public intellectuals, both criollo and peninsular. The Sociedad was one of several such groups across the Spanish empire and the globe at this time that supported agricultural improvements, economic expansion, and public education.[33] The group's primary publication, the *Memorias de la Sociedad Patriótica,* included minutes from its meetings but also a variety of articles and speeches relating to the preoccupations of the moment for the group, and for elite white Cuban society more broadly.[34]

I also draw extensively from written archival material, ranging from official proclamations, letters, and reports from individuals heading scientific projects to personal correspondence. I read all these texts both for factual information about natural history in Cuba but also for rhetorics, tropes, figures of speech, recurrent themes, and gaps. In reading these texts, I learn from scholars who have examined natural historical writing as literature to uncover metaphors and themes that this writing mobilized and the political impact of this language.[35]

Furthermore, I read the archives and texts that I consult against the grain. I attend closely to the subjects that were written about and the words that were used, as well as subjects that did not appear in this archive and terms that were avoided. I look for the excesses and repetitions of the archive that Ann Laura Stoler says serve as evidence of ambivalence and anxiety.[36] At the same time, I do not wish to overdetermine this anxiety or impose ahistorical formulations onto the narratives and concepts held by the people I was studying. I also do not wish to reimpose racial hierarchies or divisions that were and are culturally constructed. As Saidiya Hartman puts it, I do not wish to "commit further violence in my own act of narration."[37]

Each project that I examine in this book used a different rhetoric, register, and references to celebrate Cuban nature. One of the figures who stands in the background of each of these projects, however, is the German naturalist and explorer Alexander von Humboldt, whose extensive writings on Cuba in particular and the tropics in general were extremely popular across Europe throughout much of the nineteenth century. The makers of some of the projects depicting Cuban nature to these same audiences reiterated Humboldt's vision of Cuba as an undeveloped and primeval tropical paradise. Yet other members of Cuba's scientific communities focused on neatly tilled farms and plantations rather than the Humboldtian jungle. Many Cuban scientists and intellectuals spent the nineteenth century debating the proper way to administer a sugar plantation, the best crops to grow on a family farm, and the ideal land usage regulations for the island as a whole. These discussions about agriculture and land use were not abstract, but were closely tied to ongoing legal shifts and debates during this same time about the future of slavery in Cuba, what would take place after emancipation, and how to ensure that agriculture on the island survived this inevitability.

This is a book about objects and spaces, and I argue throughout that the medium and the form of these objects and spaces are central to how they work on readers and viewers. Because of this, I have divided my study roughly by medium. The first half of the book examines the spaces that were constructed in the nineteenth century for white elites to display and engage with Cuban nature. The second half examines objects—the thousands of lithographs and other prints that were produced during the mid-nineteenth century depicting Cuban nature. Many of these prints accompanied ambitious natural history texts, whereas others were created primarily for aesthetic appreciation rather than to communicate primarily scientific information.

The first chapter concentrates on Havana's botanical garden, particularly two plans for the garden that were produced in 1821 and 1828, respectively. These plans show how the emphasis of the garden shifted from a focus on scientific botany to one on agricultural botany. This shift was driven largely by the peninsular Ramón de la Sagra, who as director of the garden placed his attention entirely on non-indigenous plants that could be grown for export. These cash crops were intended to be grown in white-owned small family farms. The rising interest over the course of the 1820s in white-owned family farms reveals growing racial tensions stemming in part from the expansion of the sugar industry and of the chattel slavery that the industry required. In fact, these tensions played out on the grounds of the garden itself. The presence in the plans for the garden of quarters and provision gardens for enslaved workers reveals at least two of what Dell Upton calls "fragmentary landscapes" layered on the garden's grounds.[38] At the same time, the design of the botanical gardens allowed them to serve excellently as a space to define and perform white Cuban identity, often through contrast with these enslaved workers.

The second chapter expands this examination of public space for performing race to compare three quite different projects visualizing Cuban nature in the 1830s and 1840s. I open the chapter with an examination of how the central motif of the ceiba tree reinforced the political narratives of the newly constructed public art project El Templete. At this time of rapidly intensifying racial divisions and white anxiety, the Sociedad Patriótica turned its attention away from the botanical garden of Havana to the project of an agricultural fair. This fair changed over the course of the decade from a showcase of raw products grown on Cuban farms to a celebration of white Cuban craftsmanship. The chapter ends with a look at the natural history museum planned at the same time as the agricultural fair by the criollo naturalist Felipe Poey y Aloy. This museum was meant to be supported by citizen scientists around the island and to evoke patriotic and spiritual affect in visitors. Poey's project depended on a deeply sited, embodied local expertise that, he implied, was held only by white men, primarily criollos, who had spent nearly their entire lives in Cuba. This group included Poey himself.

My third chapter turns to printed depictions of Cuban nature. In 1835, Sagra left Cuba and moved to Paris to produce an illustrated natural history book on Cuba. The result, *Historia Física, Política y Natural de la Isla de Cuba,* celebrated professionalized science produced in the European metropole. The illustrations that accompanied Sagra's text reinforced his claims, that true science could only be performed when specimens were removed from Cuba and transported to Europe. I compare this publication with Felipe Poey's *Memorias Sobre la Historia Natural de la Isla de Cuba* (1851), which presented Cuba as an interdependent ecosystem that continued to require embodied (white) knowledge to access.

The book ends by examining several ambitious lithography projects of the mid-nineteenth century that presented divergent visions of the Cuban landscape. First, the French artist Frédéric Mialhe produced two series of lithographic prints of Cuban landscapes and cityscapes during the nineteenth century that remain cultural touchstones for the history of the island. These were *La Isla de Cuba Pintoresca,* 1839–1842, filled with picturesque landscapes populated by people of various races, and *Viaje Pintoresco al Rededor de la Isla de Cuba,* 1847–1848, which focused primarily on the streets of Havana and (often reductive stereotypes of) the people who filled them. I discuss the shift that took place between these two projects as driven partly by rising racial tensions of the 1840s. I then examine several important map and panorama projects of the mid-nineteenth century, for which artists and printers attempted to condense and render visible the changes to the city of Havana and to the Cuban landscape that defined the era. Finally, I discuss Eduardo Laplante's illustrations for the book *Los Ingenios* (1855–1857). These illustrations presented a mechanistic Cuban sugar system that would continue indefinitely into the future. These were the last major projects visualizing Cuban nature that were produced before the Ten Years' War (1868–1878), the first of Cuba's battles for independence from Spain.

1. "In Order to Dominate It"

The Botanical Garden of Havana

In 1796, a botanist living in Cuba wrote to the head of the Real Jardín Botánico in Madrid, proposing the formation of a related botanical garden in Havana, writing that "what can be attested of this island, the unexpected thing about it, is the variety and exuberant tenacity of its vegetation, different in different locations, which yield large numbers of fruits in each successive season, and which multiply during advantageous years, as well as the ability of its diverse terrains, the fertility of its waters, and the suitability of its temperate climate to support not only the growth of exotic plants but also increase the vitality of indigenous ones."[1] In the decades to come, the botanical garden of Havana, located just outside the walled city, became a built environment that was used to think through in spatial terms these geographical and climate advantages of Cuba as well as the relationship between indigenous and exotic plants and between Havana and Madrid. What made Havana's botanical garden unique, however, was how it also served as a spatial matrix for the city's white elite to think through the relationship between botany and race during a series of charged social moments. It is therefore an appropriate place to begin, as it can be thought of as a microcosm rendering tangible the conversations between science and power that drove so many of the projects in this book.

Scholars in the history of science have long noted how botanical gardens around the world served as nodes in exchange networks and as tools to extend imperial power.[2] At the same time, studies of garden design have uncovered details about how these designs changed over time.[3] There has been little scholarship, however, on how different garden designs themselves reflected the sociopolitical roles these spaces played. Havana's botanical garden did indeed serve to extend Spain's economic reach in Cuba, and its first design reflected this role, with extended visual references to the Real Jardín Botánico in Madrid. Between 1821 and 1828, though, the entire layout of Havana's botanical garden changed. Where previously geometrically intersecting paths led visitors through plants from across the island and around the world in what was intended to be a dizzying display of biodiversity, in 1828, the main section of the garden was turned into a model farm with tilled rows of familiar cash crops.

This change in the garden plan was motivated by shifting views of empire and loyalty held by Havana's white elite. During this time of growing racial unrest, the botanical garden of Havana was used explicitly to drive white immigration, imagined by white elites to quell this unrest. It became a site for white elites to redefine their own racial identity, often through the public performance of walking through the garden. At the same time, throughout the nineteenth century, the botanical garden of Havana depended on the labor and expertise of enslaved black Cubans, and the plans of the garden as well as the written archives hint at the lived experiences of these workers in the space.

"Effective and continuous communication": 1821 Plan for Havana's Botanical Garden

When Havana's garden was built, botany had been used to strengthen Spanish power across the world for many years. The reign of Carlos III, from 1759 to 1788, saw the establishment of the Real Jardín Botánico of Madrid and the Gabinete Real de Historia Natural. These institutions, in turn, funded expeditions to fill their flowerbeds and storerooms. Of course, Spain was not the only imperial power exploring and cataloging nature at this moment. As just two examples, Captain James Cook's three famous voyages to the Pacific from Britain took place from 1769 to 1778, and France's Louis-Antoine de Bougainville circumnavigated the world from 1766 to 1769. As Daniela Bleichmar argues, however, what distinguished the Spanish expeditions from other imperial scientific projects at this moment, and what makes them of particular interest to art historians, is their intense emphasis on visual evidence.[4] Members of expeditions to present-day Chile, Peru, New Grenada, New Spain, and elsewhere collected many thousands of specimens and created even more textual and visual records of plants and animals as they mapped Spanish territory in detail. The most ambitious of these was the Malaspina Expedition, which lasted from 1789 to 1794 and traveled around the entire globe, resulting in 14,000 plant specimens, 900 drawings (not including maps), and 500 zoological specimens.[5]

Cuba hosted its own share of Spanish botanical expeditions and imperial scientific interest at the end of the eighteenth century. As early as 1793, the Real Jardín Botánico of Madrid was paying a correspondent in Havana, Mariano Espinosa, to send the institution sets of Cuban flora specimens every year.[6] Then in August of 1796, Joaquín de Santa Cruz y Cárdenas, the Conde de Mopox y Jaruco, received a commission from the king of Spain for an expedition to explore Cuba. The result, the Real Comisión de Guantánamo, now known primarily as the Mopox expedition, was a set of reports on various regions of Cuba and their botanical and agricultural potential, as well as hundreds of detailed maps, plans for canals and other infrastructure, and botanical paintings. Cuba was also visited in late 1787 by the peninsular Martín Sessé y Lacasta as part of his wide-ranging Expedición Botánica al Virreinato de Nueva España. Together, the Sessé and the Mopox expeditions constituted the most complete early botanical pictures of the island of Cuba and placed the flora and fauna there squarely in the purview of the Real Jardín Botánico of Madrid. Sessé had been working on assembling a collection of plants and seeds from across Spanish territories to form the basis for a new botanical garden in Mexico City, which was to be in close contact with the Real Jardín Botánico of Madrid. He left a set of this collection in Havana when he visited, and this would form the core of the botanical garden collection in the decades to come. It was Sessé who wrote the quotation that opened this chapter, celebrating the "variety and exuberant tenacity" of Cuba's vegetation and the gains to be achieved by cultivating and distributing this vegetation.[7]

Investment and interest in Cuban botany waned slightly in the years after these expeditions until a new Intendente de Hacienda, Alejandro Ramírez y Blanco, arrived in Havana in 1816. Ramírez had a long history of interest in botany and natural history as a way to increase the profitability of a colony. He had previously been stationed in Guatemala, where he introduced a range of species including cinnamon, pepper, and Tahitian sugarcane. He had personally brought sugarcane plants back from a trip to Jamaica.[8] Upon his arrival, Ramírez set about modernizing Cuban society, especially in the fields of technology, education, and science. He took leadership of the education committee of the Sociedad Patriótica, which had previously been a small band of gentlemen amateur scientists and public intellectuals, and from this position, he used the Sociedad to drive investment in Cuban science.

Just a few months after Ramírez's arrival, and likely due to his influence, the *Memorias* published an article calling for a botanical garden in Havana. The article imagined a garden focused on indigenous flora ("the most precious plants that are produced in Cuba") which could be used for botanical study.[9] In May of 1817, the Sociedad acquired a plot of land on which to build the proposed garden.[10] By December of 1817, an article in *Memorias* reported that the Spanish crown had granted royal approval for the garden, construction of the main building had begun, and commissions had been established to collect local flora for the garden.[11] Thus, from its very beginning, Havana's garden arose in the wake of massive imperial investment in exploration and exchange and was drawn into existence by one of the highest-ranking imperial officers in the city.

An even more telling moment in the early history of Havana's botanical garden, however, took place a few years after its founding, in 1821. This was the year when Agustín Argüelles, an official in the Ministerio de la Gobernación de la Península, Sección de Fomento, in coordination with the Secretario de Despacho de la Gobernador de la Península, proposed a set of improvements to the space, which had remained somewhat underdeveloped since its foundation. Argüelles and his fellow imperial officials hoped, in particular, that Havana's central location would have enabled the garden to serve as a key node in a global botanical exchange network. In their proposal, plants were to be transferred from Spanish overseas territories to the botanical garden of Havana for cultivation and propagation, and from there sent on to Real Jardín Botánico of Madrid. As Argüelles described the proposal, it was hoped "that by putting Mexico, Manila, and other provinces in correspondence, they could exchange the plants and trees of those countries that were useful to the arts, industry trade and food of the people, and procure their propagation."[12] This would be especially important in the case of loss—transatlantic shipments of biological material were often destroyed by insects, improper care or preparation, or even shipwreck, and a shorter path for ships to sail could mitigate against some of these losses. If some of these new species became sources of profit for the crown, all the better.

This was a moment, of course, when the Spanish empire was undergoing massive changes across the globe. Many of the colonies the empire had held were being lost to the wars of independence raging across North and South America, and new sources of funds were urgently required. It was also considered important to strongly assert control over Cuba so that the island did not follow its neighbors in calling for independence. The proposal by Argüelles was intended to strengthen this control. It imagined not just a revitalized network of botanical gardens, with Cuba at its center, but a network and a colony that was firmly under the governance of the Spanish crown and whose profits flowed back to the crown.

José Antonio de la Ossa, the garden's director at the time, responded to the official request by submitting a report on the state of Havana's botanical garden on December 28, 1821. This report focused largely on the challenges that Ossa had faced for the previous four years, including the poor soil of the location chosen for the garden and his lack of funds. He expressed hope that additional financing would be provided by the crown. More funds, he argued, would in turn improve the garden, allowing for "effective and continuous communication" between Havana and Madrid.[13] In this report, then, Ossa pledged loyalty to the Spanish crown and agreed to this new proposal to position Havana as a hub of global specimen exchange to profit the Spanish empire.

Importantly, Ossa was pledging this loyalty in the wake of rapid local as well as global political changes in the governance of the Spanish empire. A crisis of sovereignty had followed Napoleon's deposition of Ferdinand VII in 1808, and the Regency Council that came into power established a representative government and wrote the liberal Constitution of Cádiz in 1812 to try to shore up support. Among other things, this constitution suspended press censorship and established narrow paths to citizenship for people of African descent for the first time.[14] However, this constitution was reversed, and new forms of rigid political and racial control were put into place when Ferdinand VII returned to the throne in 1814. These new controls especially chafed the wealthy criollo families on the island, who had gained significant political power during the period of liberal rule and resented the additional rights granted to peninsular Spaniards over themselves. Many in Cuba, of course, were acutely aware of the changes to Spain's imperial reach at the time, and although some Cubans were beginning to call for their own independence, many others were consciously positioning the island as the only truly faithful colony in the Americas to access additional political power or dispensation. As David Sartorius points out, Cuban loyalty was considered a defining characteristic of the island for these individuals, the "siempre fiel isla," as described in the title officially bestowed by Ferdinand VII in 1824.[15]

This pledge of loyalty was also rendered visible in the plan that accompanied

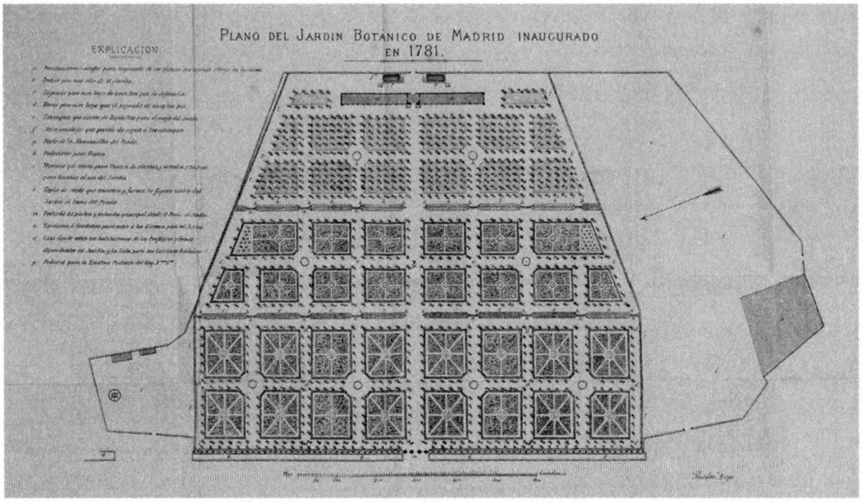

Figure 1.1. Madrid de Migas Calientes, *Plano del Jardín Botánico de Madrid en 1781,* 1875, engraving. (From *Anales de la Sociedad Española de Historia Natural* [Madrid, 1875], p. 445; Dumbarton Oaks)

Ossa's report, painted by the French engineer Francisco Lemaur (plate 1). This plan, accompanied by a detailed key, shows the rectangular shape of the garden in its location at the time inside the walled city of Havana. The entrance to the garden is at the bottom of the page and a wide central path leads away from this entrance. On either side of this central corridor a series of straight paths fork and intersect, dividing the rectangle of the garden in half vertically, horizontally into squares, and then into cross and star shapes.

This careful symmetry would have immediately called to many visitors' minds the Real Jardín Botánico of Madrid, which had been redesigned by Francisco Sabanini in 1781 (figure 1.1). Sabanini's garden included a prominent central axis with symmetrical spaces on each side that were divided by additional diagonal paths, just like Lemaur's plan for Havana's botanical garden. Furthermore, in both the Madrid garden and the Havana garden, the plants that filled these beds to either side of the central path generally increased in height toward the back of the garden. Both botanical gardens, then, seem to have been designed to provide visual access to as much of the space as possible as a visitor entered. This visitor's eye was meant to have moved straight back into the space upon entry, with orderly symmetric beds that stretched to either side, proceeding from low-lying plants to tall trees.

The long, wide central path of Lemaur's design was not unique to the Real

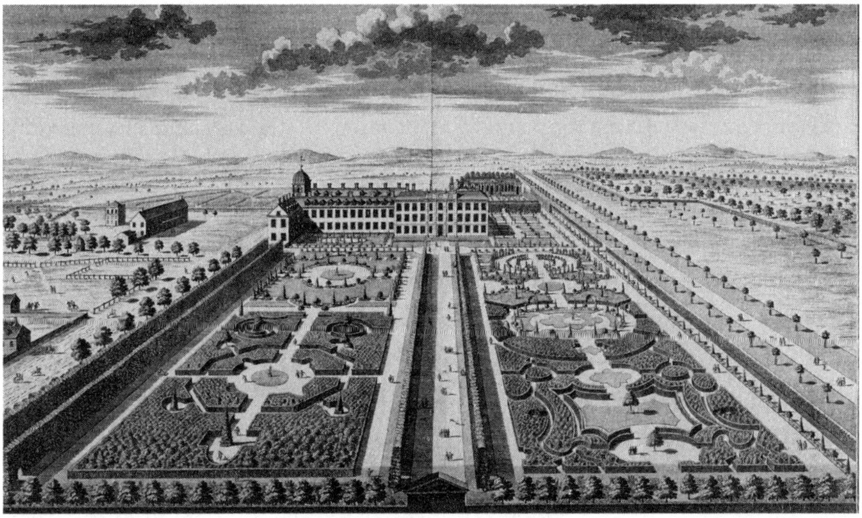

Figure 1.2. Henry Overton and J. Hoole, *Kensington Gardens, London,* circa 1720, etching and engraving. (Royal Collection Trust, London/© His Majesty King Charles III, RCIN 702920)

Jardín Botánico of Madrid, of course. Such central axes had been present in European garden designs since at least the Renaissance period in spaces such as Kensington Gardens (figure 1.2) and *Vaux-le-Vicomte.* Each of these, however, lacked the strong diagonal lines intersecting at regular intervals that defined the botanical garden of Havana and the Real Jardín Botánico of Madrid. Lemaur's plan, then, appears to have deliberately echoed the relatively recently built botanical garden at the center of the Spanish empire. This design should be understood as a visual pledge of Cuban loyalty at a time of independence movements across many other Spanish territories. This seems to have been how Spanish officials understood Ossa's report, which had been accompanied by Lemaur's plan. Notes accompanying the report show that Ossa's request for additional funds should be fulfilled to "have an effective and continuous communication between the Garden of Havana and those [botanical gardens] that are established in the peninsula" and discusses a royal order issued to Ossa to research a specific type of grass, on royal order.[16]

The design that Lemaur produced in response to the proposal from Madrid was also one of the first semi-public statements about the importance of science in Cuban life. Furthermore, Lemaur's plan for Havana's botanical garden responded to the Spanish crown's proposal with a visual statement of Cuban loyalty through botany, a statement that would have been particularly welcome at this moment of changing outlines of the Spanish empire. It was, finally, an early statement about the importance of Cuban ecology and native Cuban flora. Ossa discussed "all of the indigenous plants that had been acquired . . . with preference to those of the most interest to commerce, the arts, and sustenance."[17]

The garden under Ossa, as seen in this quotation, did contain a wide variety of plant specimens. These were arranged on Lemaur's plan to draw attention to their diversity and to reference contemporary botanical practices and research around the world. The garden sections presented by Lemaur included everything from grass, planted beds, vineyards, aquatic plants, shrubs, and beds of fruit plants. Ossa made several lists of the plants that he had successfully cultivated in the garden, which together totaled nearly one hundred species.[18] This would not have been an especially impressive selection for a botanical garden of the early nineteenth century but was still respectable, especially considering the garden had been founded only in 1817. The proposal from Argüelles, recall, was that Havana's garden collect and acclimatize plant specimens from around the world to eventually send these specimens on to Spain. Mercedes Valero González and Armando García González also mention that during these early years of the garden, Ossa published a Cuban Flora, a type of compendium text naming and giving a scientific description of all plant species known from a specific region. This book was titled *Flora Havanensis;* unfortunately, it is now lost.[19]

In examining Lemaur's garden plan and official descriptions of the role of Havana's botanical garden, it is worth remembering that the history of botanical gardens has always been entangled in the history of empire. "Physics gardens" of medicinal plants as well as a wide range of planned outdoor spaces containing plant varieties had been present across the world for many centuries, of course. Some of the earliest examples of what we now think of, though, as a botanical garden, a space for the cultivation, comparison, and study of many different species of plants, include the Royal Botanic Gardens, Kew, and the Real Jardín Botánico of Madrid. These institutions were created at the end of the eighteenth century in large part to cultivate and study the plants that were being discovered in the botanical expeditions discussed earlier that were traveling to expanding colonial spaces.[20]

Botanical gardens of various types existed in the Caribbean throughout the latter half of the eighteenth century. An inventory of the Jardín du Roi in Port-au-Prince, Saint Domingue, for example, from 1788 lists seventy-two different types of exotic trees and plants from around the globe.[21] Suriname had also been home to several attempts to found a botanical garden, in 1743 and 1787. A range of semi-private and semi-public gardens had existed in Jamaica since at least 1750, with an increase in activity in the 1770s and 1780s.[22] These early gardens of Jamaica seem to have incorporated imported plants alongside native specimens—Mark Nesbitt notes the inclusion of breadfruit plants as early as 1792.[23] Their design, however, seems to have differed significantly from that of Havana's garden, with what Jefferson Dillman refers to as "curving, naturalistic lines of plant and tree beds" rather than the strict symmetry and full visual access of Lemaur's design.[24]

Especially noteworthy in the garden's plan from 1821 is the place of pride that Lemaur gave to the Linnaean garden, immediately to the right of the main entrance. This would have been the section of the garden where closely related plants were arranged next to each other and labeled to teach visitors about the Linnaean classification system. Carl Linnaeus developed his system of taxonomy based on binomial nomenclature in his book *Systema Naturae* in 1735 and refined it in subsequent publications. The system spread quickly throughout European scientific establishments of the period and then was carried around the world through publications and botanical expeditions. Including a Linnaean garden in his plan, and specifically labeling it as such, should be understood as a deliberate choice on the part of Lemaur.[25] This emphasis on Linnaean botany in a highly visible location reaffirms the argument that the overall goal of the garden would have been the advancement of botanical research and scientific knowledge.

A celebration of local and global biodiversity had, in fact, long been acknowledged as one of the primary goals of a botanical garden in Havana. When Martín Sessé left his herbarium in Cuba in 1798, he also wrote to the Real Consulado in Madrid, stating that Cuba "should create a commission to form a botanical garden where young people studying medicine could learn, and to form a course on Botany based on what was taught in Mexico and Madrid, with examples of Cuban flora and formation of herbaria."[26] The Real Jardín correspondent Mariano Espinosa had also proposed a botanical garden in Havana as early as 1796, stating that such a garden could showcase the "variety," "exuberance," and "fecundity" of Cuba, which would in turn help the "health, economy, and commerce" of the colony.[27] Furthermore, one of the first published articles that had called for a botanical garden in Havana imagined a garden focused on indigenous flora ("the most precious plants that are produced in Cuba"), which could be used for botanical study.[28] Yet another article discussed both the emotional and the material gains to be made from Linnaean study of plants—"many times we will discover within the very country in which we live, with only a short study, equivalent plants to those which foreigners seek with their money. Above all this study of plants serves to recreate the satisfaction that generally is given to a soul when he finds that which he seeks."[29]

Despite this cacophony of praise for the idea of Cuban flora flowing through global botanical networks, at least one Spanish official expressed suspicion of these networks. This early moment in the history of Havana's botanical garden provides a kind of punctum for the idea of botanical empire more broadly. An official report about Sessé from 1796 expressed doubt about his proposal. The unnamed official who wrote the report, probably a member of the viceregal committee in New Spain, stated that Sessé was taking a house in Havana "and establishing in it another botanical garden" and then goes on to mock the very idea of a botanical garden in Havana. He wrote that if people like Sessé have their way, the crown would have not only a botanical garden in Mexico, in Havana, in the Canary Islands, and in Madrid but that "soon there will be a project to establish one in Cádiz, the Port of Santa María, Sevilla, Ecija, Córdova and other towns until there is one in the Court, to place in them and give salaries to many useless persons."[30]

This unnamed agent imagines an endlessly proliferating series of botanical gardens across the territories of the empire whose tendrils eventually grow so long and entangled that they reach back into the court itself. The horror of botanical threat reaching from the periphery back to the center is palpable. Furthermore, the most despicable aspect of these proliferating gardens, to this official, is that they did not provide utility to this empire and that they instead "give salaries to many useless persons." This quotation, though disparaging of the project, in fact accurately imagined the system of botanical exchange that was proposed by Agustín Argüelles in 1821, with Madrid at the center of a vast imperial network of plant movement. As I will discuss, however, the garden would shift in focus over the next several decades away from the Linnaean comparison of plants from around the world toward economic botany and then toward elite leisure. Before that, however, it is worth examining how race played out in this first version of Havana's garden.

In addition to having been a way to declare Cuban loyalty to the Spanish crown, the 1821 botanical garden of Havana also functioned to define and perform white Cuban identity. No records survive of visitors being specifically restricted based on race, but many public spaces were increasingly segregated at the time, as Guadalupe García explains in her excellent study *Beyond the Walled City*. The new paseos of the city, for instance, were tightly regulated in terms of conduct and race, especially at night, to ensure social control.[31] It is quite likely, then, that the garden was intended only, or at least primarily, for white visitors.

White racial paranoia was increasing during this period, driven by the memory of the Haitian Revolution and continuing rumors of imagined and actual uprisings of enslaved people in Cuba itself, most notably the Aponte Rebellion in 1812.[32] During the 1810s and 1820s, both criollo and peninsular Cubans sought to reinforce their own racial identity and privilege through a range of cultural, legal, and social initiatives. These included white-only arts academies and other cultural spaces such as the Academy of San Alejandro, founded in 1818 specifically, as Sibylle Fischer argues, to counteract the economic power of Cuban artisans and painters of color.[33]

One piece of visual evidence that the botanical garden was intended to function as a space to reify white Cuban identity is the importance given in Lemaur's plan to neoclassical sculptures throughout the grounds, including statues of Neptune and Apollo. As Paul Niell has discussed, such sculptures would become important markers of white elite identity in Havana in the decades to come, seen especially in the proliferation across the city of classically inspired fountains such as the Fuente de los Leones, Fuente de la India, and the Fuente de Neptuno.[34] Another hint of this use of the garden as a site for racial formation is the frequent use of the term "buen gusto" in the written archives connected with the garden. For instance, an early description of the founding of the garden from the *Memorias* of the Sociedad described Ramírez presiding with "buen gusto" over the choice of the land on which the garden would be located. Niell reminds us this term was often used by white elite Cubans at this time to create a narrative of cultural superiority and to distance themselves from the artistic production of free Cuban artisans of color.[35]

Although we have only these accumulated hints that the visitors to the botanical garden were almost entirely white Cubans, there is significantly more evidence that all the workers who built and maintained the garden were enslaved black Cubans. Slavery was legal in Cuba through 1886, and although the legal slave trade had ended through treaty with England in 1820, the illegal slave trade remained highly active on the island, especially during these early years. As Miguel Ángel Puig-Samper and Mercedes Valero note, Ossa wrote frequently as the director of the garden about his need for these enslaved workers because white workers could not "resist the [tropical] climate with the hoe in their hand constantly."[36] In the report Lemaur's plan accompanied, Ossa requested funds to purchase and maintain twelve enslaved laborers in total as well as two porters, a gardener, and a botanical assistant to maintain the garden.[37] These workers were needed not just for digging and weeding, but also for planting, planning, pruning, and harvesting—specialized skills that mobilized each individual's natural historical knowledge. At one point, Ossa specifically asked for individuals from the "lucumí, mandinga, or congo nations" due to what he believed to be these groups' superior strength and ability.[38]

Lemaur's plan for the garden contains two different visual references to these enslaved workers. Immediately behind the educational buildings on either side of the central axis, Lemaur depicted a symmetrical pair of living quarters for these enslaved workers. These are labeled "living spaces for the black staff." Furthermore, past the geometric beds, fountains, and statues, and a "forest of native trees for construction and other uses," lay a set of provision gardens, labeled by Lemaur as being "for the maintenance of blacks." This would have been where the garden's enslaved workers grew food to supplement the rations that were provided.

Of course, many botanical gardens, especially those in the Caribbean, Brazil, and the southern United States, would have depended on the labor and expertise of enslaved people as long as slavery remained legal in these places. Many of these workers would, indeed, have lived on the grounds of these botanical gardens and may have maintained provision gardens there as well. Lemaur's deliberate inclusion of these provision gardens in this official plan of the garden, however, remains highly unusual.[39] Even as the classical fountains of the garden likely functioned to mark the space as primarily for the enjoyment of white elite visitors, the provision garden and the path between it and the workers' quarters reinscribed black presence on the built landscape.

A white visitor to Havana's botanical garden, as envisioned by Ossa and Lemaur, then, would have strolled past these enslaved black people working in the beds and maintaining the plants. This contrast in itself, between an entirely or almost entirely white audience and black workers, would have allowed the grounds of Havana's botanical garden to function as a spatial matrix for racialized comparison and white racial identity formation. This function would have been strengthened by the presence of (also likely white marble) neoclassical sculptures scattered across this matrix and by the wide central axis of the garden, which would have functioned as a promenade and area of self-presentation.

If we understand the botanical garden of Havana in 1821, then, as a constructed geographic space, one that I argue was intended by Ossa and Lemaur to define white identity and imperial loyalty and, by doing so, to appear static, stable, and free of struggle, what was the oppositional geography of this space, according to Katherine McKittrick's formulation?[40] How did the space of the botanical garden contribute to the processes of concealment of the violence of slavery, and what lived spatial practices disrupted this concealment? Little information exists in the archives about the enslaved individuals who worked at the garden, not even their names, but the garden space itself, as shown in the plan, hints at the possibilities of resistance they held.

This racialized structuring of space recalls, for example, Maurie McInnis and Louis Nelson's examination of the reification of slavery at the University of Virginia through campus design. As they discuss, the campus was designed by Thomas Jefferson based on his own extensive experience on plantations in the American South on "how best to mask slavery by creating separate public, private, and enslaved zones in buildings and landscapes."[41] This separation of zones, of course, would have functioned quite differently than the botanical garden of Havana. In the garden, the space was designed for racial performance by public contrast—by bringing enslaved workers into the center of the garden, white visitors could have better contrasted their own leisure to this labor.

About another related space, Dell Upton discusses the embodied experience that a wide path leading up to the main house was meant to evoke for visitors to plantations in Virginia, and how this differed from the way that enslaved people who lived on the plantation themselves moved through the space. Upton describes the routes to plantations in eighteenth-century Virginia as "microlandscapes" in which "the visitor's route to the house involved passing a series of physical barriers that were also social barriers."[42] White visitors had to pass through these social barriers to be granted admittance to the great house, and the microlandscape of the plantation was designed to emphasize these barriers and create a sense of awe in the visitor. Black visitors to the plantation, on the other hand, took an entirely different route to the house, one structured not by these processional social barriers but instead by the single elemental social barrier between enslaved and free. This wide central path as a site of racializing identity might also remind us of the layout of Havana's garden, with its symmetric design surrounding a central promenade axis.

In fact, it is possible that Lemaur took the pleasure gardens around the

Figure 1.3. Detail, Eduardo Laplante, "Ingenio San José de la Angosta," 1855–1857, lithograph. (From J. G. Cantero, *Los Ingenios: Colección de Vistas de los Principales Ingenios de Azúcar de la Isla de Cuba* [Habana: Impreso en la litografía de Luis Marquier, 1857], plate 24; Beinecke Rare Book and Manuscript Library, Yale University)

main residences of some Cuban plantation owners as another model for the central axis in his garden design—as seen, for instance, in a print from *Los Ingenios,* published in 1857, which shows a garden surrounding the primary residence of Ingenio San José de la Angosta (figure 1.3). This garden bears a striking resemblance to Lemaur's plan, with a wide central path leading to a large fountain or statue, intersected by a series of diagonal paths. Much of the funding for Havana's botanical garden came from the expanding sugar industry—the Sociedad Patriótica oversaw the space, and most of the membership of this organization at the time was made up of ingenieros.

Returning to Lemaur's plan, then, we can imagine white visitors to the garden having passed through the living quarters for enslaved people, on either side of the central axis, to enter the decorative grounds beyond. This same path, however, would also have led the enslaved black workers of the garden from their quarters back to the provision gardens behind the forest after the garden gates closed. The inclusion of this provision garden on the grounds of the Havana botanical garden was a stark acknowledgment of Judith Carney's claim that such provision gardens as well as house yard gardens on plantations served as "botanical gardens of the dispossessed" in terms of the knowledge they contained.[43]

"Rich theater of scientific research": 1828 Plan for Havana's Botanical Garden

Seven years after Lemaur produced his ambitious plan for Havana's botanical garden under Ossa, a new peninsular garden director, Ramón de la Sagra, entirely reimagined the space. Under Sagra, the garden shifted from focusing on the Linnaean study of a wide variety of indigenous and imported flora to a near total dedication to economic botany and profit extracted from a small set of imported cash crops. This shift was driven in large part by the increasingly popular idea that Cuba needed more white immigrants and that these immigrants, in turn, would be attracted to better agricultural opportunities. Sagra published his plan for the botanical garden in the magazine he edited, *Anales de Ciencias, Agricultura, Comercio y Artes,* in 1828 (figure 1.4).[44]

The page in *Anales* is divided in half. On the left of the page is a plan labeled "Garden in 1827" and on the right "Garden in 1828." The page viewed as a whole, then, functions as a powerful piece of self-promotional propaganda for Sagra, who began his tenure in 1827. It presents the reader with a vision of stark contrasts and improvements under even the first few months of his directorship of the garden. The "Garden in 1827" side of the illustration represents the botanical garden (and by implication the state of Cuban science) upon Sagra's appointment as empty and overlooked. The diagram shows the rectangular plot of the garden bisected with a grid of paths, just as in the plan from Lemaur's diagram. However, where each section of this bisected grid had been filled with actual and planned plants from around Cuba and the world in the 1821 plan, Sagra represented nearly the entire garden as bare, "more of an abandoned field than a garden," as he wrote in a report to Spanish officials.[45]

This representation of Havana's botanical garden in 1827 was in direct contradiction to the plan that Lemaur had produced six years earlier, which showed the entire space filled with a wide variety of plants and carefully designed intersecting paths. The truth of the garden's state in 1827 was likely somewhere between these two representations. The 1821 plan was certainly aspirational. Ossa himself noted in the report from the same time that he did not have access to enough enslaved laborers for the conservation and improvement of the garden and that the plan represented the "continuous tasks" he had planned.[46] However, it is also clear that the botanical garden before Sagra's arrival was not the abandoned field that Sagra portrayed it as in his illustration.

José Antonio de la Ossa's report to Spanish officials had contained numerous clear requests for the resources he would need to enable Havana's garden to

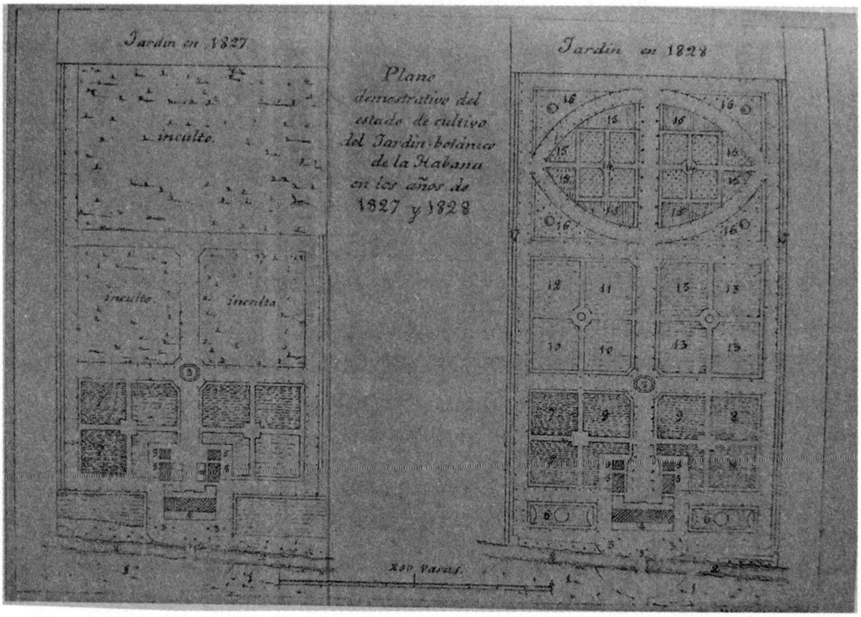

Figure 1.4. Ramón de la Sagra, "Plano Demostrativo del Estado de Cultivo del Jardín Botánico de la Havana en los Años de 1827 y 1828," 1828, engraving. (From *Anales de Ciencias, Agricultura, Comercio y Artes,* volume 2 [Havana: Oficina del Gobierno y Capitanía General, 1828]; New-York Historical Society)

serve as a major node for a transatlantic botanical exchange among Spain and her colonies. Rather than grant Ossa his requests, however, the Ministerio de Ultramar had replaced him with Ramón de la Sagra. Sagra had been born into a merchant family in Galicia, Spain, but much of his professional life was shaped by Cuba and especially by Cuban agriculture. He had studied some medicine and science at the university of Santiago de Campostela before visiting Cuba in 1821 as an assistant to a distant relative appointed to improve local tobacco production. Quickly after this first trip, Sagra got himself commissioned to be Cuba's first professor of natural history in 1823 and then appointed director of the botanic garden in 1827.[47]

Sagra wrote an effusive letter in acknowledgment of this appointment, stating that "the Island of Cuba, located under the fortunate sky of the tropics, where the plant nature is so much its strength and magnificence, can be one of the very rich theaters of scientific research, all the more so as its productions have barely been recognized." He went on to promise to grant access "to the Peninsula of the productions of this rich land."[48] This promise would have been extremely welcome to Spanish officials who faced, as discussed previously, declining imperial resources.

Sagra's plan for the garden, on the right side of his diagram, differed from Lemaur's previous plan in several key respects. In Sagra's plan, the educational center was expanded from two neighboring buildings to one large building. The northern section of the garden was changed by Sagra from diagonal intersecting lines to a complex symmetric arrangement of curved and straight lines forming an ovoid shape. Most importantly, the central section of the garden in Sagra's diagram had been divided into four sets of four quadrants. Four of these were filled with indigo, tobacco, and cotton, two with medicinal plants, one with fruit trees, one with cacao, coffee, and vanilla, and one with sugarcane. The section also included plots for germinating seeds and for additional botanical teaching. This section showcasing imported cash crops was absent from Lemaur's plan and showed Sagra's new focus on economic botany—that is, botany that emphasized agricultural improvement and increased yields of well-known crops.

In a speech in October 1824 for the public opening of a school of public botany, Sagra explained his changes to the garden's design. He stated that his proposed plan for the garden would make visible the relationships between plants of related and different families (a nod toward Linnaean scientific botany) but that the most important part of the garden (the central section) would be the section dedicated to cultivation and acclimatization. The positive results of these studies, he promised, would be visible to the public there.[49] This speech, then, was an impassioned call for economic botany. Study of crop plants and their relationship to Cuba's rich natural resources, study that took place of pride in his new garden design, would allow scientists to improve agricultural yield, better understand the relationship between different families of plants, and eventually improve the lives of Cubans across the island.

Furthermore, we can interpret these grids as visual representations of the idea that the imported crops that Sagra was growing in the botanical garden could have interchangeably substituted for each other. This "interchangeability" was echoed in a Spanish report from this same period in which a Spanish official praised Sagra's proposal for the garden. The letter celebrated in particular the benefits that this garden would offer to "peninsular industry." "There are few countries that, like Cuba, can base their industrial prosperity on a great number of products, vary them, and *substitute them with each other,*" the Spanish official wrote. "The acclimatization of these various plants may give Cuba new branches of production, thus multiplying the sources of public prosperity."[50] I believe this idea of interchangeability was a direct response to the rise of the Cuban sugar industry, which had expanded significantly in just the few years between Lemaur's plan and Sagra's plan for the garden.

This rise came about in large part thanks to the relentless advocacy of the sugar plantation owner Francisco de Arango y Parreño. In his *Discurso sobre la Agricultura de la Habana y Medios de Fomentarla* in 1792, Arango wrote that his goal was to "announce to the best of kings [the king of Spain] the opportunity and the means to give to our agriculture of the Islands advantage and preponderance over France."[51] The book called for an increase of the slave trade to Cuba as well as for investment in botany and science to expand the sugar industry on the island. In part as a result of Arango's publication, in November of 1792 the Spanish crown exempted export crops (sugar, cotton, coffee, and indigo) in Cuba from taxes and tariffs to support the sugar industry. *Discurso* was also a major contributor to increased crown support for the slave trade and rapid expansion of this trade, and as Manuel Moreno Fraginals notes, "between Jan 28 1789 and April 12, 1798, 11 royal cedulas, orders, and decrees were issued freeing and giving momentum to the great Spanish-Cuban slave trade."[52] *Discurso* also led directly to the founding of the Sociedad Económica de Amigos de País as a scientific and public service society in 1793.

Cuban sugar families formed a large proportion of the early membership of the Sociedad.[53] One of the first articles published in the *Memorias de Sociedad,* the biannual publication of notes and updates from the group, called for schools of chemistry and botany to support the Cuban sugar industry. The author evocatively described this industry as a tree itself: "The [sugar] cane is the only plant that produces a leafy branch of commerce, in whose shade most of the inhabitants of this Island live. I wish the shadow of this branch to spread through my land."[54] Another article from one of the earliest publications of the Sociedad similarly noted that "the purple bouquet of sugar turns all in its hands."[55]

Sagra appreciated sugar's profitability but remained quite concerned throughout his time on the island about the dangers of basing an entire economy on a monoculture. He wrote extensively about the need to, as mentioned above, enrich Cuban agriculture with *new* and imported crops. In a long discussion of his redesign of the garden that he published in the *Memorias* of 1824, for instance, he repeatedly invoked "the necessity of expanding the sphere of agriculture."[56] A report that Sagra wrote in 1825 also focused on his interest in "expanding the sphere of agriculture." In this report, he wrote that he wanted to support what he calls "healthy rural practices and to enrich Cuban agriculture with new and important crops."[57] Sagra included notes on a great many crops in his report, his understanding of the current state of agricultural science on these crops, and his plans for including them in the garden. He wrote about, for instance, cotton, corn and wheat, many different types of vegetables and fruit trees, and the coffee shown in one of the quadrants in his plan, noting his hope that its "cultivation will take the felicitous place in Cuban history of its greatest prosperity and wealth."[58] He also mentioned modern innovations in fertilization, grafting, and harvesting that he hoped to teach in his school of agriculture. This report, then, reiterated the fact that under Sagra, the botanical garden of Havana was reimagined as a site of agricultural experimentation and a showcase of potentially profitable cash crops to be grown on rural farms.

In Sagra's plan, the main section of the garden, directly behind the educational building, was given entirely over to agricultural experimentation with cash crops, instead of showcasing botanical diversity as it had been just a few years before under Ossa. This might seem a minor change, but it represented a major shift in how white Cuban elites understood the connection between botany, imperial loyalty, and race. Furthermore, this report points to a key moment in theorizing the intersection of botany, agriculture, race, and empire in Cuba. Sagra was particularly interested in one specific audience for his new garden designs. As he wrote in his 1825 report on the state of the garden, "when the European traveler goes through the Havana Garden School, in the shade of palms and bamboos, between rows of graceful gynogias, curious nopales, movable acacias, magnificent passion flowers, he will not be able to contain his joy contemplating the open air, the varied productions of this soil, which you will have seen in the greenhouses of Madrid, Paris or London, under glass vaults, under the heat of the heaters and among the pungent atmosphere of tannin."[59]

In other words, while indigenous plants of Cuba were of interest to Ossa because he wished to send them to Madrid and establish the island as a loyal node in the imperial botanical exchange network, indigenous plants of Cuba were of interest to Sagra because of their exoticism in the eyes of (white) European visitors. He imagined their "joy" to see plants growing out in the open that, in Europe, could only be grown under greenhouse glass. This quote shows Sagra imagining the garden as a space for the establishment and performance of white identity through affect—joy—and physical immersion in the native flora of Cuba. Similarly, Sagra's speech at the public opening of a school of public botany from the year reiterated his interest in the link between white identity and the botanical garden. "The only thing that will assure the prosperity of the island of Cuba," he wrote, "is the establishment of small crops, crops that are not grown in anticipation of exorbitant capital." More important, he went on to say that such diversified agricultural botany will "increase the necessary population of the country."[60] This phrase, "necessary population," would have been understood by Sagra's audience at the time to specifically mean the white population of Cuba.

I discussed in the introduction the dozens of articles published during the 1830s and 1840s on the urgent necessity to increase white immigration to Cuba. Throughout the first half of the nineteenth century, many white people argued in articles like those quoted, as well as in the streets and the home, that white immigration to Cuba was a crucial goal. Several laws were passed to encourage this immigration. As Adriana Chira discusses, as early as 1817, a census of the island found that the white population of Cuba was in the minority, and this result caused the governor at the time, José Cienfuegos, to call for official royal support of a white migration scheme, especially in light of free Cubans of color claiming increased political rights.[61] In October of 1817, as a result of such calls, Spain issued a royal *cédula,* or royal order, calling for an "augmentation" of the white population of Cuba. The *cédula* discusses the importance of augmenting this population with peninsular Spaniards or immigrants from the Canary Islands as well as those from Catholic European counties. This is necessary, it argues, "for the prosperity and preservation of this priceless island."[62] This *cédula* and accompanying land reform laws enacted during the same decade granted white immigrants a small plot of land exempt from taxes for fifteen years in addition to granting Cuban citizenship and ensured that white colonists would have access to land that had previously belonged to the Spanish crown. This *cédula* in 1817 followed a very similar order issued in 1815 that encouraged Spanish people, and later all white people, to settle in Puerto Rico and specifically permitted these immigrants to bring a set number of enslaved people to the island.[63]

The first half of the nineteenth century was also marked by the formation of social and political organizations to encourage white immigration, such as the Junta de Población Blanca, a committee of the Sociedad Patriótica that was first formed in 1817 but was especially popular and active during the 1830s and 1840s. The Junta's primary concern was to conduct a census of the island regularly, monitor population data by race, and suggest strategies to increase the white population in Cuba.[64] Cuba was, of course, not the only Latin American site with white immigration schemes in the mid-nineteenth century. The 1820s and 1830s also saw white artists and politicians in Jamaica seeking to, as Kay Dian Kriz puts it, refine "the rude and toxic West Indian landscape into a space capable of attracting and maintaining a healthy, virtuous, and polite white population."[65] Ernesto Bassi has written on similar schemes in Colombia in the 1820s, and Erika Edwards has recently drawn attention to racial self-fashioning in the whitening rhetoric of nineteenth-century Argentina.[66] One of the impetuses, furthermore, for calls for a white migration scheme in Cuba was the perceived success of the scheme established in Puerto Rico in 1815 in response to the aforementioned *cédula*.[67] What was unique about Cuba, however, was not just that these schemes took place under continuing imperial control rather than new nationhood, but also that they intersected with a rise in interest on the part of white elites in botany and natural history.

Many of the new regulations specifically discussed agriculture and the small family farms that Sagra imagined in his garden—for instance, a decree issued in 1819 to "promote agriculture and the white population," which specified exemption from taxes for any white settler who created a farm to cultivate cotton, coffee, or indigo—the very plants that Sagra was experimenting with and demonstrating in his garden quadrants.[68] There was also the article quoted in the introduction to this book that said, "The white population in the countryside of our island is one of the most urgent necessities." The bulk of this article discussed the cultivation of mulberry and silk, one of the possible cash crops that its unnamed author said could be grown by white immigrants.[69] One interpretation of Sagra's new design for Havana's botanical garden is that it functioned as a space to imagine these small white-owned family farms into existence, as well as to cultivate the crops they would need to become profitable.

This rise of white interest in immigration schemes must be read in part as a reaction to the continued increase of the enslaved black population of the island (despite the slave trade being officially illegal since 1820) as the sugar plantation system expanded and white anxieties about slave revolt. As Sibylle Fischer has written, the fear of the Haitian Revolution recurring on the sugar plantations of Cuba can be considered nearly an obsession on the part of plantation owners and other white Cubans.[70] Cuban officials were constantly looking for, and claiming to uncover, plans for slave revolt on plantations and in cities across the island, further stoking this fear. Ada Ferrer writes about how the idea of Haiti was used throughout the nineteenth century by Cuban planters to increase their power over enslaved people and to reinforce social control more broadly.[71] White immigrants were thought by white Cubans to counteract this possibility of slave revolt, though the exact mechanism by which this was meant to happen remained somewhat opaque. Relatedly, the Cuban sugar economy itself had expanded significantly with the arrival of white plantation owners from Haiti during and immediately after the revolution on the nearby island, often accompanied by workers that they kept enslaved.[72]

Another historical development that drove white anxieties during the 1820s and 1830s was the increasing political and economic power of free Cubans of color, especially in the city of Havana. In part because of the strict stratification of Cuban society and the disenfranchisement of almost all Cubans of color from positions in government, associations and societies were central to life in Havana. African-born enslaved Cubans and their descendants (including free blacks) in the city gathered in Abakuá lodges and *cabildos de nación,* which served as social networks, economic safety nets, spiritual communities, and performing groups. Free Cubans of color also belonged to militias as well as artisan guilds and other professional organizations. Many of these organizations also served as political advocacy groups. As David Sartorius and others have discussed, black militias and *cabildos de nación* petitioned government officials, issued public declarations of loyalty to Spain, and instituted professional regulations for guild members.[73]

During the time that the botanical garden of Havana was being built and then redesigned, then, the black and brown population of Cuba was growing quickly and members of these communities were establishing their own networks of care and political advocacy. These changes served as a significant source of racialized anxiety on the part of many white middle-class and elite Cubans. White immigration schemes and a countryside filled with small, white-owned family farms were imagined to counteract this anxiety. The botanical garden of Havana under Sagra was meant to render these farms visible, to train these white immigrants (as well as Cubans themselves) in their maintenance, and to cultivate the crops grown on them.

Sagra did have some success with his agricultural experiments. In 1830, he was issued a royal order to continue studying the possibility of growing indigo in Cuba, and in 1833 he was commended by the Intendente of Havana for his success in doing so.[74] In 1834 the Intendente of Havana sent another laudatory report on Sagra's work to Madrid, mentioning his success in creating and introducing agricultural improvements for rural areas, his successful production of indigo and mulberry trees, as well as other plants including types of gum and textiles.[75] However, the fact that none of the cash crops listed in Sagra's diagram ever made up a significant portion of Cuban exports, and the fact that sugar exports continued growing each year, should remind us that any success he did have was never replicated on a large scale. In the end, it turned out that the crops that Sagra celebrated in his grids were neither as easy to grow nor as interchangeable as he had represented them. Each of these crops required specialized equipment and knowledge, which Sagra did not fully possess and was not able to transmit to the few students who did attend classes at his garden.

Sagra, however, seems to have been more successful in his related goal, of fostering a sense of "joy" among white visitors to the garden. It was likely partly to encourage these white visitors and to foster this joy that the white elite entertainment complexes that Ana Amigo discusses were established immediately next to the garden. One of these was a theater called The Diorama, which was installed in the garden in 1827, decorated by the French artist Jean-Baptiste Vermay, and hosted dances and plays. These included a masked dance during the Havana carnival in 1831 and the dance in honor of the *jura* of the infanta Maria Isabella Luisa in October 1833.[76] In part because of these events, Amigo argues that the botanical garden of Havana became during the 1830s a site of white elite leisure and sociality. She adds that this was particularly important in a city with limited public leisure spaces. This interest in space for white elite leisure is visible in Sagra's plans for the garden itself as well, which included fountains and statues along the main corridor, where visitors would promenade up and down.

Just as Ossa had, though, Sagra also depended on enslaved workers to build and maintain the garden that he imagined in his plan. The *Memorias de la Real Sociedad Económica* from the period includes several mentions of extended negotiations between Sagra and the Sociedad regarding the number of enslaved workers that would be necessary to execute and maintain his plan for the garden.[77] The matter was finally resolved when a member of the Sociedad "donated" an unspecified number of "*negros bozales*" (enslaved people who had been born in Africa as opposed to in Cuba).[78] It is unclear if these individuals were in addition to the ten enslaved black workers who were assigned to the garden by the Captain General in 1824 that Sagra mentions in his 1825 report.[79]

Just as Lemaur's plan for the garden did, Sagra's plan also shows the quarters for enslaved laborers located immediately behind the building that served as an educational center and home for the garden's head professor (Sagra himself). Camilla Townsend discusses this proximity as a "sudden exposure" of Sagra to the realities of chattel slavery, which partially drove Sagra's lifelong suspicion of the sugar economy and slave societies more generally.[80] Unlike in Lemaur's plan, however, Sagra did not include the provision garden that these enslaved people would have used for their personal sustenance in his overall design for the garden. The provision grounds almost certainly still would have existed, but they had been excised from Sagra's celebratory public rendition of the space.

There are only a few other hints in the written archives about the experience of the enslaved people who lived next door to Sagra, or about the experiences of any non-white, non-elite visitors to the garden under his leadership. One document outlining the reorganization of garden staffing under Sagra noted that he "will be in charge of all of the slaves and field hands of the garden and will treat them with moderation and good order, he will take care to watch for any faults to avoid punishment. . . . The religious acts of the slaves of the garden are the charge of the gardener, they should take mass each Sunday and name one of themselves to go to the church."[81] It is impossible to know if this intentional outlining of proper treatment of enslaved workers was a result of conflicts and excessive punishment under Ossa or of the wider debates over the treatment of enslaved labor forces that were taking place across Cuba at this moment.[82]

Overall, however, it seems likely that in many ways, the experience of an enslaved laborer living and working at Havana's botanical garden would have changed little between 1821 and 1828. The laborer still would have worked in the garden, planting, propagating, and pruning throughout the day, and walked the long path back to the provision grounds at the end of the day, and slept in a building behind the educational complex. The laborer would have taken mass on Sunday and may have attended a church in town. Although the neat squares of crops on Sagra's plan were meant to attract white immigrants to Cuba, these plots would still have been maintained almost entirely by enslaved black workers.

"Too inviting to resist": Botanical Garden's Move to Molinos, 1839–1840

Sagra left Cuba in 1835 to travel to France and the United States to work on his natural history publication that I discuss in Chapter 3. By the end of the 1830s, the gardens he had redesigned were in some disarray. Furthermore, the leadership of the garden was being pressured to move, as the space it occupied was needed to expand a new railway station and storage depot. The proposed new location for the botanical garden, the Quinta de los Molinos, lay slightly to the north of the walled city of Havana. The move was somewhat controversial as, since its founding, the garden had been located inside the city walls, in a somewhat central location. A letter in the *Diario de la Habana* from 1839 discussed the plan indignantly:

> That the botanical garden, transferred to El Molinos, would produce the same advantages as it does at its current location seems impossible to me, because who will be the poor man who, in the force of the heat and despite his laziness, has to go to look for the medicinal plants that help his ailments and that are currently two steps from his door; [who are] those who are convalescing who go there to recover their strength; [who are] those Botany students who come to hear the lessons of their professor; who are the farmers who busily ask about their crops, visit the establishment, examine their wealth and acquire inspiration and knowledge to guide them in the new branches of rural industry? . . . Where would the owners of plantations get the products they need? For example, the bread tree, without which I would have had to spend a lot of money last year on the maintenance of my blacks.[83]

This quotation neatly summarizes the shifting goals of Havana's botanical garden in the previous decades—to be a source of medicinal plants, information on crops, and the development of rural industry. It implies that at least some of Sagra's goals were recognized as successful to at least some of his audience in Havana, including the letter writer, who did see the garden as a source of information about agricultural innovation and crops. The letter also reminds us of how the botanical garden, and agricultural science more broadly, supported and reinforced chattel slavery—as mentioned in the last section about the bread tree saving money on the "maintenance of blacks." Furthermore, as the letter hinted, none of these functions would have happened as effectively if the garden were moved to the outskirts of the city. This proved to be prophetic. The gardens under Ossa had been a site for celebrating biodiversity and under Sagra a site for imagining agricultural alternatives to the sugar industry, but in its new location the botanical garden functioned almost exclusively as a pleasure garden for Havana's elite. Joseph John Gurney in 1840 called it "a place of constant public resort."[84]

The movement of the garden to this new space was overseen by Miguel Tacón, who served as Captain General of Cuba from 1834 to 1838. Under Tacón, the lived space of Havana changed drastically. He oversaw the construction of wide, lush pedestrian walkways such as the Paseo de Martí reaching through the city, planted with trees for shade. He also drove the construction of a series of infrastructure projects, such as an aqueduct, a railroad, and a massive new prison, as well as cultural centers such as the Tacón Theater. Beginning in the mid-1830s, then, new public spaces were being created for the use of Cubans, and there was increased public discussion about the appropriate use of these spaces.

The new location for the botanical garden was directly adjacent to a large villa for the Captain General that Tacón had also overseen the construction of. Because of this, many visitors know the space as the "Tacón Garden." Thus, the head of the Spanish government in Cuba, as part of a general program to redesign the city of Havana, had moved the city's botanical garden to the space directly next to his new villa, and the garden became known by this same government official's name. We might read this as a reinscribing of imperial power on the spaces of Cuban botany, an echo of Argüelles's proposals for drawing Cuba's botanical garden more firmly into the networks of Spanish botanical empire.

This shift in focus was made evident and was reinforced by the new layout for the garden, which James Philippo calls "gardens laid out in the style of Versailles."[85] One visitor in 1845 described the experience of the gardens at Quinta de los Molinos in detail:

> We'll follow the crowd to the Tacón Garden, some very prettily laid-out grounds enclosing the summer residence of the captain governor [. . .] a beautiful moonlight, we sauntered down an allée, that looked too inviting to resist; we pursued it till we found ourselves in a sweet garden, where we wandered wondering and delighted. At every turn a new and agreeable surprise; a jet d'eau; a gurgling waterfall, with its moss and grottos; we ascended terraces, sat down in arbors, wound through thick-leaved groves [. . .] I had no idea where we were, whether on public or private property. [. . .] I thought of the magical gardens in the Arabian Night.[86]

Similarly, Samuel Hazard wrote in 1873 that the gardens "are open day and night, and any one is allowed to enter and stroll through the beautiful walks, shaded and surrounded by most exquisite tropical flowers, shrubs, and trees."[87]

The earliest plan that survives from the Quinta de los Molinos location is

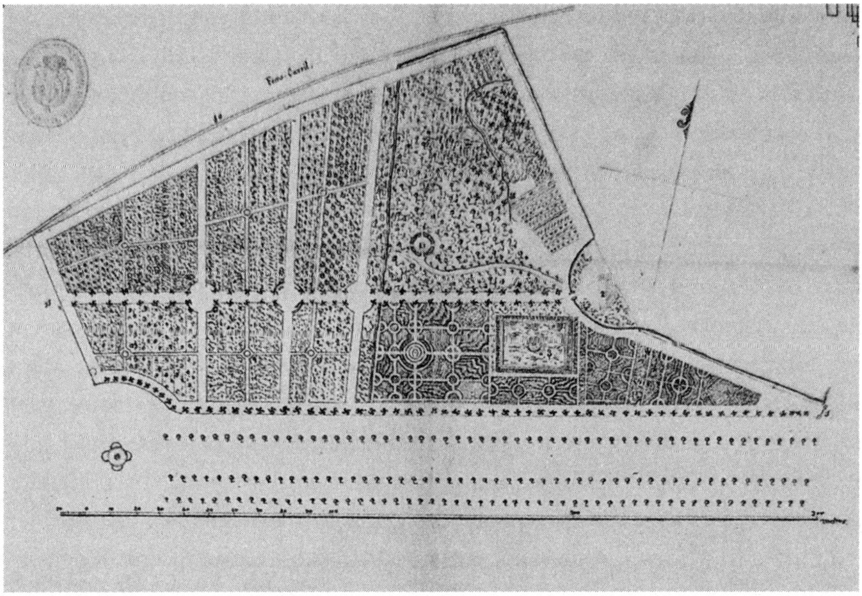

Figure 1.5. *Plano Topográfico del Jardín Botánico de la Havana,* 1864, engraving. (Archivo Histórico Nacional, Spain)

from 1864 (figure 1.5). This plan shows the irregular shape of the space divided by a long central path, with a series of paths intersecting it at perpendicular angles. This design, then, seems to reference the earlier layout of the garden in its previous location, with its wide central axis and symmetric paths. However, the above description of the garden seems to refer specifically to the upper right quadrant, in which the orderly beds of Ossa's and Sagra's gardens have been replaced by a "ramble" with sinuous paths snaking through the space.

Ian McLean and other art historians have discussed the apparently random but actually highly structured and planned sinuous path of a garden ramble as "an aesthetic of colonial identity" in the context of British colonization of Australia, which functioned in part by rendering familiar the unfamiliar landscape of the colony.[88] Whether or not this is how the paths of the garden at Quinta de los Molinos functioned, what is apparent from the descriptions above is that the garden had been stripped of nearly all its scientific associations in the minds of visitors, and had become primarily a site of leisure and retreat. The travelers that visited described wandering the grounds, enjoying the atmosphere, and appreciating the flora for its aesthetic qualities, but never mentioned any kind of intellectual engagement with the botany of the site. There are no records of the kind of agricultural classes or scientific experiments taking place in the Quinta de los Molinos location that had been cornerstones of the garden in its original location under Ossa and Sagra. And, notably, both the Linnaean garden and the agricultural experimentation zones had been excised in the plan for the garden at its Quinta de los Molinos location.

What is also apparent from the quoted description of visiting the garden is that the space had become inextricably linked in the minds of visitors to the Captain General himself. When the gardens were moved to Quinta de los Molinos, they were understood by the visitors no longer as a public botanical garden but as a private garden belonging to the Captain General. Thus, the garden that white elite criollos and peninsulares alike had used for several decades to assert loyalty to Spain, but also to imagine different forms of local nature, in its new location was entirely subsumed under the assumed ownership of the highest-placed local Spanish official. Although Amigo makes a convincing argument that the botanical garden served as a space to imagine alternative modes of citizenship and sociability beyond the strictures of the colonial agenda, in the end it remains unlikely that to international visitors at least this space of possibility was claimed by the empire itself.

None of the designs that survive of the gardens in their Quinta de los Molinos location include living quarters or provision grounds for the use of the workers who maintained the grounds. The visual record that would teach us about the experience of these workers, therefore, is non-existent. The only hint in the national archives of Spain and Cuba about the experience of the black men and women who worked at the new location of the botanical garden is in a set of letters from 1864 and 1865 that refer to eight freed black workers at the garden and requests for funds to hire an additional eight freed black workers.[89] There is no indication of why emancipated black laborers specifically were desired, given that slavery was still legal in Cuba through 1886. It is possible that some of these workers at the garden were the same individuals who had been enslaved by the Sociedad in the decades prior, individuals who had purchased their freedom or had it purchased for them. And the Sociedad may have wished to continue employing them in the garden because of their expertise and familiarity with the plants.

These archival puncta, then, raise more questions than they answer. Regardless, these mentions of black workers should remind us that throughout its history the botanical garden of Havana depended not just on the labor but also on the natural history knowledge of the black workers, free and enslaved, who worked there. These individuals maintained the grounds as the focus and the layout of these grounds shifted significantly. They would have laboriously dug up each of the plants and helped move them across town, to the grounds next to the Captain General's house to create the "beautiful walks" and "sweet gardens" described by visitors.

Throughout the nineteenth century, the way that the botanical garden of Havana represented Cuban nature to the various audiences who visited the space changed significantly. Havana's botanical garden in the 1820s under Ossa responded to a proposal from Madrid to be a center of exchange for a newly revitalized network of imperial botanical exchange with visual statements of loyalty and an emphasis on Linnaean botany and gathering many different types of specimens. Sagra reorganized the garden in the 1830s to focus on economic botany, specifically interchangeable crops to be grown for profit by small, white-owned family farms. Under both of these directors, the garden also provided a space for white elite visitors to perform racial identities in contrast with the enslaved black workers who maintained the grounds. This performance of white identity, especially by leisurely strolling through nature, became the primary goal of the garden in the 1850s in the Quinta de los Molinos location, one which was also intimately connected to imperial power.

These changes, from a focus on Cuban biodiversity to a focus on imported crops for small white-owned family farms to a focus on white elite leisure, are reflected in the shifting designs for the grounds of the garden. Finally, while little archival evidence remains of the experiences of the enslaved black workers whose labor and expertise built and maintained the garden, a few hints of these experiences do allow us to imagine the different experiential landscapes that make up the diagrams and plots that we see today. These diagrams remind us of the enslaved workers who transplanted botanical specimens, maintained their quarters, and tended their provision gardens in the matrix of the botanical garden. We can extend this examination of differentiated experiential landscapes and the racial performance of semi-public space to other locations in Havana. This is what I will discuss in depth in the next chapter.

2. “The Peculiar Circumstances of Cuba”

Public Space, Museums, and Exhibitions

The 1840s in Cuba were defined by rising racial tensions, which ignited in 1844 with massive public trials accusing black and brown Cubans of plotting an island-wide slave uprising. These accusations, in turn, were met with extensive state violence against the accused and entire reams of new legal codes to restrict the rights of Cubans of color. These rising racial tensions overlapped with, and often shaped and were shaped by, substantial changes to the urban fabric of the city of Havana. Together, these forces led to the creation of new public and semi-public spaces, where identity and race were policed.

Just as had been the case in the botanical garden, many other public performances of white identity took place in highly planned outdoor spaces and used the metaphor of Cuban nature to think through the changing social systems of the period and to further define white Cuban identity. In this chapter, I begin by discussing racial identity formation in the new public spaces of Tacón's Havana more broadly and then examine two spaces in depth—a planned natural history museum and an agricultural fair. The natural history museum of Havana was intended to create and reify a new criollo citizen science. The agricultural fair, in turn, celebrated white craftsmanship and small white-owned family farms, just as Ramón de la Sagra's farm laboratories at the botanical garden of Havana had. Both spaces helped to define white identity and to propose a future for Cuban nature outside of the expanding but increasingly fraught sugar plantation system. I end this chapter with a brief examination of Cuba's

Figure 2.1. Jean Jacottet, after James Gay Sawkins, *Vista de la Plaza de Armas,* 1839, engraving. (Harvard Art Museums/Fogg Museum, Gift of Charles H. Taylor, Photo © President and Fellows of Harvard College, M3494)

displays at the international expositions of the second half of the nineteenth century. These displays crystallized many of the associations between productive Cuban nature and desired white immigration to the island that drove the earlier projects.

The process of public racial identity formation within the changing urban fabric of Havana is rendered visible in the *Vista de la Plaza de Armas,* a lithograph by the British artist James Gay Sawkins printed in Paris in 1839 (figure 2.1). This print is drawn as if the artist were standing on the steps of El Templete, a new structure that had been commissioned by the Sociedad just a few years earlier. The plaza itself had been in place for more than two hundred years but had recently been renovated under the leadership of Captain General Miguel Tacón.

Sawkins's view deemphasizes the buildings that surround the plaza and fills the garden at the center of it with greenery, which had been absent in previous prints showing the same plaza. Most importantly, he placed significant visual emphasis on the paths through the wooded park, expanding these paths and filling them with figures of various races and classes strolling and carrying goods. These figures are depicted speaking in pairs and looking at the other people in the space, as in the two white men at the center of the print who look at an approaching black man carrying a large piece of lumber. This is an image of urban circulation and interaction, but it highlights the mixed racial makeup of the population. This view depicts Cubans defining themselves racially in relation to one another and constructing separate racial identities, using the new public spaces of Havana as a stage to do so.

As discussed in the previous chapter, Havana was reimagined under Captain General Tacón in the 1820s and 1830s. He oversaw the construction of public promenades and parks throughout the city as well as the new railroad, prison, and theater. As Guadalupe García has argued, many of these changes sought to regulate race in the city of Havana, either by restricting the movements and the locations of black Habaneros or by producing a hyper-visibility that enabled social control.[1] The new spaces that Tacón helped create also often combined nature and public performance of identity, such as the new paseos and the renovated parks planted with lush greenery and filled with Cubans of different races promenading. These public spaces also enabled additional surveillance, both from official police forces and from fellow Cubans.

It is this public performance and surveillance that is at the heart of James Gay Sawkins's lithograph of the Plaza de Armas. Furthermore, just as had likely been the case at the botanical garden, this performance was structured and reinforced by native and imported Cuban plants. The trees, shrubs, and flowers that filled the paseos and parks were crucial to the function of these spaces. Greenery provided shade and visual interest, giving Cubans a reason to promenade in the first place. Neat rows of trees fill the center and the edges of Sawkins's print, and the figures in the print walk among them.

This intersection of public racial identity performance, new public spaces, and Cuban nature also drove the creation of El Templete, which functioned as the vantage point of the Sawkins lithograph. As the name suggests, this was a small, temple-shaped public monument, visible on the right side of Frédéric Mialhe's lithograph of the Plaza de Armas (figure 2.2). It was commissioned by the Sociedad and built in 1827 quite near the Captain General's palace. The monument still stands in the same location. Consisting of a single rather small room, the structure could be entered, though access was often blocked by a gate that could be locked. In the years after El Templete was built, its interior was covered with painted murals depicting moments from Cuba's history.

Paul Niell has discussed the construction of El Templete as a complex statement of criollo identity that brought together the classical architecture of

Figure 2.2. Frédéric Mialhe, "Plaza de Armas (Habana)," 1839–1842, lithograph. (From *La Isla de Cuba Pintoresco,* 1839–1842; University of Miami Library, Cuban Heritage Collection)

the building itself with a specific narrative of Cuba's past, made through site and imagery.[2] As Cuban-born whites, criollos were often granted less political power than the peninsulares (white people born in Spain) on the island who occupied all of the highest political positions in colonial society. An aspect of this culturally significant public artwork that has remained underexamined, however, is the relationship between El Templete and wider discourses of Cuban nature. Much of the imagery that fills the interior of the space, as well as its external decorations, revolves around trees, specifically the ceiba, or silk cotton, tree. El Templete was built at the site of an ancient ceiba tree that was said to be where Spanish conquistadors founded the city, conducted the first mass, and held the first meeting of the *cabildo* (town council) in 1519. This was the same ceiba tree where, according to an ordinance of 1573, any African slave caught carrying weapons was to be publicly lashed. By the time El Templete was inaugurated, the actual tree was long gone, replaced with a carved pillar representing it. The monument, then, used the symbol of a remembered historical tree to construct a Cuba defined by an idealized past and maintained through violence enacted on black bodies.

The original ceiba lived again, in fact, in the three massive history paintings that covered the inside walls of the monument. These were painted by

Figure 2.3. Jean-Baptiste Vermay, *The First Mass,* circa 1827–1828, oil on canvas. (Oficina del Historiador de la Ciudad, Havana)

Jean-Baptiste Vermay, a French painter who also had been serving as the first director at the new Academy of Art of Havana. The three paintings, completed some months after El Templete was opened to the public, depict the first *cabildo* (plate 2), the first mass (figure 2.3), and the inauguration of El Templete itself, this last painting filled with the figures of the contemporary Habanero citizens who had attended. Taken together, according to Paul Niell, these paintings narrate an "allegory of progression from the primitive space of trees, Indians, conquistadores, and virgin landscape to a highly populated nineteenth-century city with its stone ceiba tree monuments, architecture, and refined and well-clothed residents."[3]

The ceiba tree itself acts as the main character in each of these scenes. Vermay depicted the first mass and the first *cabildo* with the tree towering overhead, a thick brown trunk behind the central figures in each scene, and a dense net of green leaves filling nearly the top third of each massive canvas. In the third canvas, of the inauguration, this canopy is gone, replaced by an expanse of blue sky. This sensation of enclosure and then openness adds to the narrative of progress that Niell attributes to these paintings. The monument as a whole used the symbolism of the tree to construct a specific history and local identity for Havana at a time of rapid change. The rhetorics of historic continuity and unique criollo identity put forth in the history paintings commissioned by the Sociedad were, crucially, linked to a specifically Cuban species of tree, and to a specific tree growing in Cuban soil itself.

Furthermore, El Templete served as a site of public identity formation concerning the historic scenes that it included. The viewer who had been granted access through the gate was meant to stand under the painted ceiba tree and identify with the (peninsular and criollo) figures depicted in the scenes, looking up at the same tree. This viewer might also have looked around the small space at the handful of other visitors, almost certainly all white, all also identifying with the historical figures in the painting. These visitors would have then exited El Templete to stroll the paths of the Plaza de Armas that surrounded the structure, reinforcing this historically informed white Cuban identity by comparing themselves to the multiracial crowds that fill the paths between the plaza's trees and plants, as seen in the Sawkins print.

"Assembly of the patriotic collection": Building a Natural History Museum in Havana

Soon after the cityscape of Havana had been redesigned under Tacón, a young scientist named Felipe Poey turned his attention to the project of creating a new natural history museum for the reimagined city. Poey spent over a decade assembling a collection of thousands of specimens, commissioning display cases, and negotiating for a location for the museum. This project not only represented an important moment of advancement of natural history in Cuba in the mid-nineteenth century more broadly, but also a specific vision of natural history space driven by criollo citizen science at a time of increasing calls for Cuban independence.

There had been several museum spaces and semi-public scientific collections in Cuba from the very beginning of the nineteenth century, most notably an anatomical cabinet with wax demonstration figures created by the Sociedad Patriótica around 1820.[4] The idea for a natural history museum, specifically, on the island was likely first proposed in 1835 by José María Zamora, an economist and legal scholar who was serving as director of the Sociedad at the time. Little is known about what Zamora had planned or accomplished, only that it was meant to be a "cabinet of natural history" and that he worked for a brief period with Sagra on the project, which was then abandoned.[5] Felipe Poey was appointed director of the society a few years after Zamora's proposal and restarted the project in 1838.

Poey had been born in Cuba—his family owned sugar plantations near Matanzas. In 1838, he was best known for having written a geography of Cuba for use in primary schools as well as an illustrated natural history about the butterflies of Cuba.[6] By the time he turned to his museum, Poey had spent years thinking through how best to represent the full extent and complexity of the island of Cuba to both general local audiences and specialized natural history readers across the world. Poey was also, however, quite consciously and deliberately a part of transatlantic natural history networks, having studied natural history in Paris for years as a young man, and his work on Havana's museum reflected his experiences in Europe as well. In 1826 he brought the eminent French naturalist Georges Cuvier a set of eighty-one drawings of Cuban fish and a barrel filled with eighty-five specimens of Cuban fish.[7] He also brought Cuban plant and animal specimens to additional scientists and organizations in Paris, including the Jardín des Plantes.

Poey planned for his museum to bring together these two aspects of his scientific engagement—embodied expertise with Cuban nature and insistence on the place of this embodied expertise in the most refined European natural history sites. Throughout his planning of the museum space, he also imagined it as an inspirational site that would create a sense of pride in visitors, especially fellow criollo visitors. He wrote that "a museum of this nature, presenting to the youth of Havana a series of indigenous and exotic productions, conserved with cleanliness and order, and accompanied by instructive cards, will attract the eyes and then the spirits to the study and greater knowledge of said objects."[8] Poey's vision of the museum, then, sought to attract "the spirits" of visitors to the study of the homeland.

Although Poey's museum did not open to the public until much later in the nineteenth century, a report he published in the *Memorias* of the Sociedad in November of 1840 clearly outlined his collecting practices and his vision for the space. By this time, Poey had amassed a significant collection of specimens, divided into what he called the "Cuban museum" and the "exotic museum." The Cuban museum held "some" mammals, thirty birds with their nests, the eggs of fourteen species of birds, fifty fish, various reptiles, three hundred shells of marine mollusks, one hundred terrestrial snail shells, two thousand insects, an herbarium of five hundred plants, a collection of seeds, a set of sixty varieties of woods, and two hundred minerals. The exotic museum held some additional mammals, twelve birds, three hundred marine shells and one hundred terrestrial shells, fifteen hundred insects, two hundred plants, a marble inlaid table, several Native American artifacts, and a library of sixty volumes of natural history books. It also included a tusk of an elephant and one of a rhinoceros, a live crocodile and a live dog, and two (presumably dead) "monkeys from Africa."[9] Assembled in just a few years, the entire collection totaled more than five thousand specimens.

Poey went on to write that he had been enjoined by a French naturalist, Sauveur Abel Aubert Petit de la Saussaye, to "start to buy the furniture, stands, boxes, drawers, knobs . . . they are necessary to facilitate classification; without them, you do not know where to focus your attention, because it's like being in a labyrinth."[10] The cases and stands that would display specimens were therefore considered crucial to organize the very information being conveyed in the museum space, much like margins on a page or dividing lines in a chart. Once these cabinets were acquired (a process that would, in fact, prove to be quite difficult), "skeletons will be forms for the comparative study of anatomy, taking care to put [together] in the cabinet one species of each family or general principle."[11]

Poey's natural history museum, then, was meant to have been a temple to Cuban and global biodiversity, in which systematic comparison led to increased scientific knowledge. This vision accords closely with the current form of the museum. Still open to the public on the campus of the Universidad de la Habana, the single-room space contains large cases made of wood and glass that hold dense displays of specimens from Cuba and abroad, arranged according to anatomical similarities to allow for visual comparison. Poey's plan for the space also accorded closely with many of the major natural history museums of the period, such as the grand gallery of the Muséum d'Histoire Naturelle in Paris (figure 2.4), in which similar specimens were placed next to each other in glass-front cases to allow for anatomical comparison.

Throughout his career, Poey remained highly aware of the centrality of Paris to global academic natural history networks. Paris was home to the botanical garden, natural history museum, and learned societies that were held in the highest esteem by almost anyone involved in academic science in the nineteenth

Figure 2.4. J.-B. Pujoulx, "La Grande Galerie du Muséum d'Histoire Naturelle de Paris," 1803, engraving. (From J.-B. Pujoulx, *Promenades au Jardín des Plantes à la Ménagerie et dans les Galeries du Muséum d'Histoire Naturelle,* volume 2 [Paris: Librairie Économique, 1803]; Bibliothèque nationale de France, département Sciences et techniques, S-33203)

century.[12] As for Poey, in addition to having studied there, he was in regular communication himself with many scientists in Paris. Poey's use of the natural history museum in Paris as a reference for his own museum in Havana, as well as his quotation of a French naturalist in justifying the expense of the display cabinets required to members of the Sociedad, show him cannily using the reputation of French natural history to his own ends. It is somewhat ironic, then, that Poey turned the reputation of French natural history to the creation of a uniquely Cuban form of natural history.

In fact, in the same report in *Memorias,* Poey goes on to quote Petit de la Saussaye at great length, not only on the planned design of the museum but on the importance that the project would hold for Cuba.

> I am pleased to be able to say that finally the island of Cuba will be in possession of a cabinet of natural history, a worthwhile endeavor which will contribute greatly to the progress of natural history, seeing the richness of the land and what little is known of it. No one more than you [Poey] could undertake such a project, because you hold for it the sacred fire which is the greatest guarantee of correctness . . . [expeditions to the interior of Cuba will build the museum's collection], seeking in its forests, its mountains, its rivers and in its ports and coasts the natural products

> that serve not just to enrich the patriotic museum, but also to get via exchange exotic products.[13]

By quoting a French naturalist at length in his report to fellow educated members of the Sociedad Patriótica, Poey not only enhanced the legitimacy of his project, but also reinforced his idea of the museum as a "patriotic" endeavor. It would seek in the forests, mountains, rivers, ports, and coasts of Cuba the collections that would demonstrate the richness of the land. Poey described the museum repeatedly, in fact, as a "patriotic collection" and a "patriotic museum."[14]

This was a moment when the Spanish empire was still seeking to reassert its power in the remaining colonies, including Cuba, in large part because, as mentioned in the previous chapter, many of its previous colonies were being lost to the wars of independence raging across North and South America. By 1823, this reassertion of control led to the liberal Cadíz constitution being nullified in Cuba and significantly increased Spanish military presence on the island. Spain ruled Cuba without a constitution until 1837 when a new centralized government imposed a set of special restrictive laws. These restrictions included increased censorship of Cuban presses and heightened trade restrictions. Many criollos had a sense of political exclusion and felt a growing discontent during this period.[15] Nature metaphors were often used at this time to discuss the rise of independence movements across Cuba. An article in the *Memorias* of the Sociedad Patriótica from 1823, for example, blamed these independence movements on "the dark agents of other emancipated provinces [who] sow the fatal *seeds* that inflame the imagination of the young drugged youth."[16]

At least one scholar argues that Poey was one of these youths, stating that the reason Poey had gone to Europe in the first place was that his ties to activists for independence landed him in legal trouble in Cuba in 1822.[17] Poey indeed complained to a friend this same year about having more government "restrictions than France currently has" and mentioned the "new blood spreading through the veins" of Cuba with the possibility of national representation.[18] For decades after his fiery youth, Poey remained closely associated with many of the strongest advocates of Cuban independence, including Félix Varela y Morales and José de la Luz y Caballero. Rosa María González López has pointed out that Poey also publicly opposed the Spanish crown on at least one occasion, the expulsion of the British consul from Cuba in 1842.[19] It was quite likely, then, that throughout the time that he was planning his museum, Poey harbored nationalist sympathies of his own.

At the same time, Poey also professed loyalty to the Spanish crown and its local representatives. Poey petitioned the Captain General at the time, Jerónimo Valdés, for additional funds for his museum. He then went on to laud the Captain General's generosity and reiterate his own fealty, writing in *Memorias* that "such a noble and generous demonstration of interest from our illustrious leader serves the progress and advancement of the country."[20] Overall it seems that Poey worked continuously to navigate the increasingly treacherous waters of political discourse in Cuba in the 1840s.

In this context, his use of the word "patriotic" in describing his museum should draw our attention. The word was not, of course, unusual at this time and likely did not have the same connotations of support for independence that it does to the modern reader. Recall that the Sociedad itself was using the name the Royal Patriotic Society of Havana throughout the 1830s. The term, however, does hint toward the criollo pride that I believe was one of Poey's driving motivators in planning his museum of Cuban and exotic collections.

Poey's rhetoric repeatedly centered affect as a driving force for the project. He discussed the "most intense ardor" that ought to drive botanical study.[21] He also called the museum "a moral establishment, since in it the study of natural beings raises the soul towards knowledge of God, the creator of these marvels of the world, and descending pleasantly the youth of vain pastimes, direct it to study and contemplation, worthy use of time, free from all remorse."[22] Poey's conception of natural history as a path toward divine inspiration and intense emotional experience was unusual for Cuban natural history discourse of the period. None of the documents associated with the botanical garden and very few of the articles in Cuban scientific journals of the time cited such reasons for studying botany. Furthermore, none of Poey's own previous writing on natural history used such language. It seems that it was the natural history museum space specifically that Poey imagined as capable of inspiring such affect. Another way that Poey imagined his museum as capable of inspiring patriotism, emotion, and pride was through collecting practices, what I call a Cuban criollo citizen science.

Poey insisted, throughout his long article from 1840 on the museum as well as in many other reports on the progress of his project, that the majority of the museum's collections should have been not just indigenous to Cuba but also collected by ordinary Cubans instead of professional scientists, from the "streets and in the fish markets."[23] Poey first proposed this "citizen science" approach in 1838, reporting that he had begun "correspondence with the necessary professors from the island and abroad, as well as had [invited] the general public to respond to the opportunity" to contribute to the museum.[24] He placed an announcement inviting these contributions in the daily newspaper *Diario de la Habana* as well as in the monthly magazine *El Plantel*.[25] Another article in the *Diario* about this crowd-sourced collecting plan echoed Poey's language, calling the plan "interesting and patriotic."[26]

This type of citizen science was unusual at the time. One of the defining qualities of botany in the first half of the nineteenth century, particularly in Europe, was its increased professionalization. Dorinda Outram, for example, argues that the gentlemen amateur scientists of the late eighteenth century were replaced in the early nineteenth century in Britain and France by expert researchers working in state-funded and state-controlled institutions such as the Natural History Department of the British Museum, the Zoological Museum of London, and the Muséum d'Histoire Naturelle in Paris.[27] At the same time, however, a different model of scientific production was being developed in the new republic of the United States based almost entirely on observations reported by ordinary citizens. Andrew Lewis has written that the authority given to eyewitness accounts in the United States at this time, in particular accounts of animal behavior, reflected a belief in the unique qualities of American nature as well as the democratic ideals of the new nation. According to Lewis, this scientific authority given to individuals living in the United States would open science "to all Americans" and would reveal truth through a "competition between facts."[28] This is also how Poey's vision of citizen science functioned—to elevate ordinary Cubans and give them the authority to define the unique nature of the island.

Since Poey had studied science in Paris for several years and was also in close communication with scientific societies in London and New York, he would have been extremely well informed about these various models of botany at this moment. His choice, therefore, to recreate the model of citizen science from the United States over the model of state-sponsored professionalized science from Europe would have been a deliberate one. And Poey's museum did, in fact, help further knowledge of Cuban nature. For instance, he oversaw several expeditions to the island's interior to supply the museum with specimens. The *Memorias* reported on one of these expeditions in September of 1839 that resulted in new minerals and other specimens for the museum and another in April of 1840 in search of "insects and other objects of natural history."[29] Furthermore, the donors of the specimens that Poey listed in his 1840 article truly had come from across the island and even around the world. These donors included a Cuban nobleman living in Paris, a writer, a country priest, a doctor, and a sugar mill owner.[30] As this list makes clear, these individuals were not quite the "fishermen" and "general public" that Poey had originally imagined would send specimens. They did, however, represent a wider cross-section of Cuban society than had been involved in natural history projects on the island up until this point. As the list also implies, all of these donors were white men, and almost all had been born in Cuba.

Poey's decision to exclude black and brown Cubans from his call for contributions to his museum can be contrasted with several white scientists of the era who did recognize black scientific knowledge. The British naturalist Philip Henry Gosse, for example, listed Richard Hill, a Jamaican scientist of mixed European, African, and East Indian ancestry, as co-author of *A Naturalists' Sojourn in Jamaica,* and Michel Étienne Descourtilz cited medical knowledge held by black women in Haiti in his books.[31] I believe that Poey's exclusion of Cubans of color from his museum project can be partially explained by the rising racial tensions in Cuba at the time. As Guadalupe García discusses, the Bando de gobernación y policía de la Isla de Cuba (Proclamation of government and police of the Island of Cuba) in 1842 represented a crucial moment in these racial tensions. This set of government regulations covered a wide range of aspects of the daily life of Cubans, including health, hygiene, and public morality. Many of the regulations sought to reestablish racial boundaries and police the livelihood and movements of black and brown Cubans. Owners of lodging houses, for instance, had to provide local police with a list of their lodgers every night, and anyone renting rooms to black Cubans was specially instructed to check all documents to ensure they were not forgeries. Similarly, employers "had to register the names, addresses, and legal status of the black men and women who used the [public] promenades, along with the number of the volante (carriage) in which they drove or sat, or else suffer a fine of twenty pesos."[32] Given these restrictions across the city, it is likely that Poey intended to use the space of the museum to claim knowledge of natural history for white Cubans, especially criollos, to increase the number of white Cubans who would study the field. The natural history museum, thus, was meant to be a space of racial self-definition.

To reiterate, Poey's plan for the museum was a large space filled with a series of wood and glass cabinets, each containing specimens encouraging visitors to study and compare them. This space would also have contained large curiosities, such as the canoe mentioned in his 1840 article. This was a space that would have been open to the public, meant to "attract the eyes and then

Figure 2.5. Charles Willson Peale, *The Artist in His Museum,* 1822, oil on canvas, $103\frac{3}{4} \times 79\frac{7}{8}$ in. (263.5 × 202.9 cm). (Courtesy of the Pennsylvania Academy of the Fine Arts, Philadelphia; Gift of Mrs. Sarah Harrison [The Joseph Harrison, Jr. Collection], 1878.1.2)

the spirits to the study and greater knowledge" of the natural history objects on display. White visitors would likely have seen other white visitors in the space and identified themselves as a racial unit, surrounded by displays celebrating Cuban flora and fauna.

This white identity, in turn, would be linked to the "patriotism" that Poey repeatedly cited in his writing on the museum. Poey's use of massive curiosities such as the canoe and the tusks to anchor the space and inspire patriotism can be compared to Charles Willson Peale's museum in Philadelphia (figure 2.5). Peale's museum, founded in 1784, juxtaposed cases of natural history specimens with portraits of American political leaders and massive curiosities, such as a skeleton of a mastodon. Peale's museum space, like Poey's, was intended to foster wonder, patriotism, and an appreciation for the richness of local nature.[33] In Peale's case, this was patriotism for a newly founded nation. In Poey's, it was patriotism for a unique criollo identity that may even have pushed against Spanish imperial claims to the island.

Despite his success in collecting specimens and leading expeditions, Poey's museum was plagued by funding shortages and seemingly intractable challenges to finding an appropriate space for his planned displays. By 1845, the Sociedad was losing patience with Poey's delays and appointed a pair of individuals to look into the situation and determine what would be necessary to open the museum to the public.[34] The commission, and Poey himself, reported back to the Sociedad several times over the coming years about the continuing financial difficulties faced by the institution and the increasing likelihood that it would not open to the public. The problems with the museum grew so overwhelming that by 1849, a decision was made to officially transfer all of the museum's holdings to the Real Universidad in Seville.[35] The collection, which Poey had envisioned as a patriotic display of Cuban flora and fauna driven entirely by citizen scientists and deeply tied to the island itself, was thus absorbed by the imperial metropole. Little information is recorded about Poey's feelings about this absorption, only the following note in the *Memorias* of 1849: "Remembering the impossibility of sustaining the establishment [of the museum] at the time being . . . and desiring the organization contribute by all possible means to the scientific advancements of the country and knowing that the museum of the Royal University is somewhat lacking, [Poey] agreed to surrender the objects of natural history [to Seville] with the promise that the collection be returned if at some time the Society needed them.[36]

I do not believe it is a stretch to read ambiguity and regret into this description of Poey's transfer of his hard-won collection. At the same time, however, the natural history knowledge that Poey gathered during his preparation to open the museum was put into use in two books on Cuban nature that he wrote in the coming years, as will be seen in the following chapter. Furthermore, the museum did open several decades later and, as mentioned, to this day follows the general plan he laid out for the space and includes some of the specimens that Poey prepared.

"An effective media to promote the advance of this noble art": Agricultural Exposition of 1847

The halls of Poey's museum were meant to celebrate Cuban whiteness, especially uniquely criollo identity, within the frame of Cuban natural history. During these same years, years of increasing racial tensions and ever louder calls for independence, another display space was created. This space, the agricultural exposition of Havana, was also meant to celebrate "ordinary citizens" of Cuba, specifically the white small-farm landowner that Sagra had tried to call into existence fifteen years earlier at the Havana botanical garden. Over the course of the planning process for this exposition, from 1843 to 1848, the focus of the agricultural fair organized by the Sociedad Patriótica shifted from a livestock exposition to a celebration of industry, manufacture, and craft.

The first time that an agricultural exposition was discussed by the Sociedad was in a long proposal published by its Committee of Agriculture and Statistics in the *Memorias* of November 1843. The planned fair was to be held at Havana's hippodrome, another addition to the city that had been made under Captain General Miguel Tacón. The proposal for this exhibition centered on livestock and animals. Livestock, the report explains, "form a very important part of our riches, and will soon be called upon to nourish our rural laborers."[37] Almost all of the proposed prizes were for livestock. These included prizes for the best bull, the fattest ox, and the cow that produced the best milk. In addition to livestock, however, "there should also be prizes for sugar and coffee, as well as tobacco, corn, vegetables, other products."[38]

The idea of an agricultural fair, particularly one that focused on livestock, was not, of course, unique to Cuba. Such events were quite common at the time, especially in British colonies, including Australia and Canada. They took their form from the British country fair and, as several scholars have recently traced, were instrumental in spreading agricultural knowledge and improved production techniques across wide swaths of colonies and former colonies.[39] Such fairs were also quite popular in the United States. These events were often organized by agricultural improvement societies and other groups of elites, similar to the white elites who made up the Sociedad. The Philadelphia Society for Promoting Agriculture, for instance, organized a number of these events.[40] The Sociedad, then, was borrowing the form of the agricultural fair from other locations, likely to help establish Cuba as a technologically modern agricultural society. In fact, in the first discussion of the fair in the *Memorias,* the author of the article talks about these agricultural fairs held outside Cuba at length. "The committee is aware that in all the towns where agriculture flourishes, they have established agricultural fairs, courses, and livestock exhibitions, as an effective media to promote the advance of this noble art, the base of the prosperity of the region."[41] The article goes on to discuss England, in particular, as a site of great advances in agriculture and animal husbandry and states that the country's agricultural fairs served as a major driver of these advances.

The Sociedad's decision to mount an agricultural fair in Cuba in the 1840s, however, also stemmed from the group's ever increasing preoccupations with the island's demographics and with the intersection of race and rural agriculture. These preoccupations played out most clearly in the activity of the group known as the Junta de Población Blanca (Committee on White Population). First formed in 1817 as a committee of the Sociedad to promote white immigration, the Junta was primarily a surveying body. It was responsible for producing detailed reports on agriculture and population data across the island with the largely unspoken assumption that it was only by knowing these details that white Cubans could be successful in shifting the demographics of Cuba back toward a white majority.

One of the primary projects undertaken by the Junta de Población Blanca during the 1830s was an agricultural census of the island listing each plantation and farm, surveying which crops each grew and how many enslaved and free laborers worked on each. Reports from the group on the progress of this census appeared in nearly every issue of *Memorias* during 1838 and 1839. The Junta followed this agricultural census with an island-wide population census conducted in 1841, the first such census conducted in nearly fifteen years and only the second census conducted in the history of the Sociedad.[42] This is all to say that the idea of mounting an agricultural fair coincided with a moment of extreme preoccupation on the part of the white elites of the Sociedad with the racial makeup of Cuba, and especially with the racial makeup of rural agricultural production. This was the same moment as the dozens of articles that were published in the *Memorias* about the possibilities of "plantations without slaves," the need for white immigration to the island, and also the need to create small family farms to entice and occupy these (still largely imagined) white immigrants.

White migration was not mentioned explicitly in the article on the planned exhibition, but this context was a clear subtext of the entire proposal. It was visible, for example, in the use of words with racial connotations throughout the proposal such as the "casta" and "raza" of the cattle.[43] This language reinforced the white supremacist logic of white Havana in the 1840s, in which race and caste were the defining qualities of existence. Geographic determinism was also quite present in this proposal, which discusses the natural resources of the island and the "propio pais" (their own country) of the products.[44] This celebration of the natural fecundity of Cuba's soil, of indigenous Cuban nature, should remind us of the "Cuban collection" of Poey's natural history museum. We can assume the exhibition organizers also hoped that this natural fecundity would attract white immigrants to the island.

Furthermore, from the very first discussion of the exposition in *Memorias,* the event was meant "to recognize the work and intelligence of the rural laborers who work unrecognized" and to inspire future innovation. The article goes on to say that "these men honored and publicly stimulated by the awards and certificates of the Agriculture Section would strive to be better, would furthermore have this comfort amid so much fatigue."[45] These "rural laborers" who were repeatedly cited throughout the description of the planned exhibition should be understood as the white owners of small family farms. If the article were referring to the enslaved laborers on plantations, its author almost certainly would have used the term "negros" or "esclavos"—as, for instance, in the series of articles in the 1830s on challenges to acquiring more enslaved workers, "brazos esclavos," for plantations.[46] The exhibition in its original planned form, then, seems to have sought to elevate rural laborers and the small white-owned family farm to encourage white immigration to Cuba. This was quite similar to the reformulation of Havana's botanical garden under Sagra as an exhibition of agricultural technologies and cash crops that could be grown on these farms. The garden also likely served as a public space for the performance of white identity through comparison with black laborers in the space.

There are, in fact, some indications that racial performance was also one of the goals of the Sociedad's Agricultural Exhibition. The article in *Memorias* ended with a discussion of how the exhibition might have intersected with Cuba's plantation complex, considered by many members of the Sociedad to be the reason for the increasing black population that they sought to mitigate. "When the exposition moves to [other parts] of the country," wrote the author, "the Committee [of Agriculture and Statistics] can choose in anticipation the most appropriate site without much trouble since the heads of ingenios, ranches, coffee plantations, and other smaller farms will offer their possessions to contribute to this way of realizing the patriotic work of the group."[47] Various plantations would have, therefore, hosted versions of the exposition as a patriotic gesture celebrating the productivity of Cuba. White visitors to these exhibitions could have imagined the white family farms the displays celebrated and compared this ideal to the enslaved black labor of the surrounding plantation space. The plantations—the primary source of racial anxiety for white Cubans—would thus also have driven the reduction of this anxiety.

The generalized racial anxiety discussed previously was often projected by white Cubans onto the plantation itself and a corresponding preoccupation with the possibility of slave uprising. Both the Haitian Revolution and the Aponte Rebellion of 1812 still felt quite recent to planters in the early 1840s. At the time of the planned exposition, many plantation owners were expressing frustration with Spain's failure to protect private property (including enslaved workers) and worries about interventions by British abolitionists and U.S. annexationists. A hurricane also hit Havana in September of 1842, causing significant destruction, and enslaved workers rose up in large numbers, likely to strategically take advantage of the political unrest of the period, in March and November of 1843.[48] These were, of course, not the first slave revolts or the first rumors of independence and annexation. These uprisings, however, added fuel to the fires of paranoia of slave revolt among the white elites of Havana.

This paranoia continued to blaze when the summer of 1844 was defined by drought across Cuba, and in October of the same year, when another powerful hurricane hit western Cuba, destroying vast swaths of the plantations there as well as causing significant damage to the urban fabric of Matanzas, Cardenas, and Havana.[49] Then, in December of 1844, a planter in Matanzas claimed to have discovered evidence of a massive planned slave revolt. This claim truly exploded the previous decades of racial tension and racial paranoia. The Captain General immediately announced military rule on the island and initiated an investigation across the entire island. The officials who conducted this investigation soon claimed to have exposed an even wider conspiracy that involved hundreds of activists for independence, prominent white foreigners living in Cuba, and well-known free Cuban intellectuals of African descent such as the poet Plácido.

The degree to which this widespread conspiracy actually existed remains a matter of debate among historians.[50] What is clear, however, is that government reprisal for the perceived revolt was swift, all-encompassing, and terrifying. Within a year, many thousands of free and enslaved people of color "had been executed, banished, imprisoned, or had simply disappeared" in what became known as the Year of the Lash.[51] Particularly targeted were the other professional, social, and spiritual organizations of Cubans of color. Many of these organizations were disbanded, and onerous regulations were placed on those that remained. Free black Cubans were no longer allowed to serve in militias, and many laws were enacted restricting their occupation, their conduct in public, and their social activities (all of which had been, as we have seen, highly regulated even before this moment).[52] The events of 1844 affected every member of Cuban society, rewriting and then violently reinforcing the racial divisions that had already been present.

This important series of events was, somewhat surprisingly, never mentioned in any of the files surrounding the planning of the agricultural exposition, or in the *Memorias* itself. In the year after the discovery of the supposed conspiracy, however, the *Memorias* insisted ever more urgently on the need to increase the white population of Cuba. Reports on that topic appeared in every issue from July through November of 1845. Regarding the exposition itself, a government report in February 1845 stated that it had been postponed simply due to the hurricane.[53] The event was delayed several additional times, for unspecified reasons, and finally took place in November 1847. Even though the events of the Year of the Lash were never discussed in relation to the agricultural exposition, the exponentially intensified racial paranoia of white elite Cubans at the time was rendered visible in the final form of even this small and seemingly unracialized exposition space.

Although the proposal from the fall of 1844 emphasized livestock competitions, the exposition that actually took place in 1847 instead seems to have consisted almost entirely of crafted and manufactured objects that had been produced in Cuba. A report on the final form of the first exposition in the *Memorias* of 1848 listed the types of objects that had been on display:

1st: all kinds of carpentry, masonry, cabinetmaking, blacksmithing, pottery, locksmithing, shoemaking, boilermaking, silverware, tinsmith, watchmaking, cordmaking, tobacco shop, and hair hat manufacturing

2nd: Of fine fabrics, those that are made of linen, cotton, yarn or silk, and embroidery

3rd: Of common fabrics those of filaments of bananas, lacebark tree, ficus, pines or any other similar matter; and the construction of straw hats and mats

4th: Thick fabrics, *serones* [woven straw bags], sacks, *jabas* [a different type of woven straw bags], baskets, *nasas* [woven fish traps], *chinchorros* [a different type of woven fish trap], hammocks

5th: The compounds of the vegetable kingdom such as chocolate, bread, noodles, wine and spirits of different fruits, preserves of these fruits in syrup or any other kind of delicacies, and also the best dried fruits made for export, and finally the most varied and new sweets

6th: Also vegetable products that are the processed and purified oils of coconut, palm, sesame, and of all species of seeds; cassava starch or other roots, gums, resins, balms extracted from the trees of the country, oils and flower essences, colors for the art of painting or dyeing, and the manufacture of white, brown and painted paper

7th: Of the animal kingdom, the tanned skins of animals, and their transformation into the various known or new classes, for common use and other arts, soap, salted or brine meats and fish, cheeses, butter, hams, and all kinds of preservation, and finally all the machines, industry, arts, and chemical products that are considered useful will be objects of the exhibition.[54]

This list made clear that the only objects that were displayed in the 1847 exposition were those that had been crafted or processed in some way.

The exposition of 1847 took place in the building owned by the Sociedad. This building was not large—certainly, it was much smaller than the hippodrome that had originally been proposed for the exposition. Very little information survives on how the objects listed above were laid out in the space of the Sociedad's building, which had a small courtyard at its center surrounded by wide halls. Exhibition designs during this period ranged from items placed haphazardly on tables to massive halls filled with elaborate booths and carefully constructed displays. It seems, however, that Havana's exposition fell somewhere on the former end of the spectrum, according to an account by the governor of the time. "People accustomed to the magnificent spectacle offered by public exhibitions in foreign countries, where thousands of objects made with admirable care and perfection, dazzled their sight, captivating their understanding at the same time, will have found this first public exhibition of Cuban industry quite limited and even paltry."[55]

This description shows that the members of the Sociedad were quite aware of exposition practices in other countries and were seeking to emulate them. Of course, it also makes clear how, even to a supporter of the Sociedad writing in its own publication, the exposition fell short of these lofty goals. Based on this description, as well as the list of objects on display at the exposition, we can imagine the final form of the event as a series of tables set up inside the (somewhat dark and cramped) Sociedad headquarters, with the crafted objects listed laid on top of the tables. The list of objects includes fabrics and tanned animal skins, which may have been hung on walls or spread out on tables. It also includes larger crafted objects—tools and pots (produced by blacksmithing), cabinets and furniture (under carpentry and cabinetmaking), and hammocks and baskets. These large domestic objects would then have been joined by a wide range of foodstuffs—chocolate, bread, wine, fruit preserves, dried fruit, jars of oils, resins, balms, cheeses, and preserved meats. Finally, the display included processed materials to be turned into additional crafted objects—processed plant fibers, paints, and paper. Again, taken together, this list consists entirely of objects for domestic use that were produced on a small scale, sometimes by a single individual. The objects listed, mostly small-scale domestic items like pottery and fabrics, would, I believe, have made the interior space of the building feel almost like the small white-owned farm the exposition sought to valorize.

The shift in the focus of the exposition to handicrafts and small-scale manufacturing is especially surprising as, up until this moment, Cuba had not been particularly known for such objects. Most written accounts of the island at the time did not mention manufactured or crafted objects at all and instead focused on the island's natural wealth, its fecundity, and the diversity of its flora and fauna. The governor of Cuba went on to say that "the most thoughtful visitors [to the event] will attend to the peculiar circumstances of Cuba which is essentially agricultural, as well as to the novelty of an institution that requires familiarization with our manufacturers and craftsmen."[56] This shift to "manufacturers and craftsmen," then, was acknowledged as "novel" and requiring "familiarization."

The inclusion of crafted objects into an exposition planned by the Sociedad is also surprising because the Sociedad had not previously shown any particular interest in handicrafts. They are never mentioned in *Memorias,* for instance. And perhaps most importantly, an exposition planned by the Sociedad focusing on crafts is surprising because the majority of artisans in mid-nineteenth-century Havana were not white. By the first half of the nineteenth century, Havana contained an extensive and highly developed system of craft guilds of free artists of color, guilds that included furniture makers, weavers, and, most famously, the mural painters of Havana.[57] Free black mural painters were responsible for decorating the inside and outside of many of the most elaborate houses in Havana in the first decades of the nineteenth century, and these homes would then have been filled with furniture and fabrics created by other artists of color in the city. These guilds had, in fact, been the loci of racial anxiety on the part of white elites for decades.

As Sibylle Fischer argues, it was in large part to counteract the popularity of black and brown mural artists and other artisans that the Sociedad Patriótica founded a new drawing academy in 1818. Fischer goes on to say that attendance at this new academy was almost certainly limited to white Cubans, particularly given the passionate calls in the months leading up to its founding that admission to the academy be available only to students who were "white, of known parents, and [with] a good education."[58] Despite the presence of the academy, white elite Cubans remained highly concerned about the power, wealth, and prestige of black and brown artisans. The "problem" of black artists was addressed again in José Antonio Saco's jeremiad from 1831, "The Arts Are in the Hands of the People of Color." Saco wrote that "among the enormous evils that this [African] race has brought to our land is that they have alienated our white population from the arts. . . . In this deplorable situation, no white Cuban could be expected to devote himself to the arts, because the mere fact of embracing them was taken to mean that he renounced the privileges of his class."[59]

The criollo intellectual Félix Varela y Morales also decried the presence of artisans of color. "Virtually all free people of color are engaged in the skilled crafts," he complained. "For each white artisan, there are twenty of color."[60] Brown Cuban artisan guilds were some of the organizations that were highly suppressed during the Year of the Lash. As Michele Reid-Vazquez notes, the military commission trials that followed the Year of the Lash banned many Cubans of color from working as artisans and oversaw the deportation of thousands of prisoners, many of whom had also made their living as artisans of various kinds. Furthermore, one of the most publicized trials of the Year of the Lash was that of the poet Plácido, who also worked as a "silversmith, printer, comb maker, and carpenter."[61] The final form of the Sociedad's exposition, then, took place not only after a moment of intense racialized violence but also after decades of insistent white opposition to artists of color.

White elites had long expressed deep discomfort with the black and brown artisans of Havana. The Sociedad had previously sought to undermine these artisans, in particular by founding a new academy of drawing. These artisans of color were persecuted singly and in guilds during the racist violence of the Year of the Lash. Though the scanty archives that survive from planning the 1847 exposition do not mention the race of the participants in the event, we can assume, given the above evidence, that it was not these artisans of color who were invited to participate in the event organized by the Sociedad. It was extremely likely that the exposition instead showcased the few white artisans who were working in Cuba.

I believe that this exposition was intended specifically to further strip black and brown artisans of Havana of their market and to establish white artisans in their place. Again, many Cuban elites were deeply invested in ensuring that white people outnumbered people of color in the ranks of Cuban artisans, especially during and immediately after the Year of the Lash. Thousands of artisans of color were deported to North Africa and Spain or semi-voluntarily emigrated to Mexico and elsewhere in the Americas during the military reprisals of that year. In large part because of these forced movements, the Sociedad Económica de Amigos de País of Matanzas, to the west of Havana, proudly reported a 67 percent increase in the number of white artisans in the city between 1846 and 1849.[62]

If this segregation theory was true, this exposition would have functioned similarly to the botanical garden and the natural history museum of Havana. All three spaces would have been used by white visitors to perform whiteness, by being in the space together, recognizing each other, and engaging in leisure activities together. The exposition, further, would have enabled this white visitor to have imagined a white-owned family farm filled with the domestic objects on display, somewhat like how Sagra's experimental beds at the botanical garden also invited visions of this same white-owned family farm. Finally, the makers themselves at the exposition would likely have been white, allowing the visitors to the space to have removed black and brown Cubans entirely from this reverie of the domestic agricultural future of Havana.

We should note, furthermore, that displaying locally produced handicrafts and manufactured objects may also have had proto-nationalistic implications at this moment of growing calls for Cuban independence. Although Spanish economic policy toward Cuba shifted significantly over the first half of the nineteenth century, many merchants and some planters agitated during this period for increased access to free trade with countries other than Spain.[63] The Sociedad's exposition was a celebration of objects made in Cuba, for Cubans—effectively removing Spanish imports, and perhaps by extension Spain itself, from this vision of the domestic space as well. This reading of the exposition is complicated, of course, by the fact that the Sociedad's politically diverse membership was composed at this time of several extreme loyalists as well as some of the figures calling most urgently for independence. It remains noteworthy, however, that this was a display of locally produced objects at a time of increased imperial presence and debates about tariffs and restrictions on trade.

"The merit they need": Cuba at International Expositions

The agricultural exposition of 1847 was important not just because it likely represented a vision of a white rural Cuba in the years immediately following extreme racial violence, however. The experiences of the Sociedad in planning the exposition also shaped how the group decided to represent Cuba to international audiences at global expositions throughout the rest of the nineteenth century. Cuba participated in at least five major international expositions during these decades, and it was almost always the Sociedad itself, or leaders of the organization, who planned and funded these displays. These displays generally alternated between celebrations of Cuban craftsmanship and presentations of Cuban raw materials that painted the island as a site of natural fecundity and biodiversity. This shift back to biodiversity is seen, for instance, in the fact that Cuban nature was featured in Spain's display at the Universal Exposition of 1851 in London in a set of Cuban wood samples collected and arranged by none other than Ramón de la Sagra.[64]

This collection of wood samples had been assembled by the Sociedad beginning twenty years earlier, in 1831. Sagra, then serving as director of the botanical garden, was the head of the committee producing the collection. As such, he oversaw requesting samples from Sociedad members, organizing these samples, and preparing them for display. The final samples were cut to the same size, polished, and placed together into drawers. Each was labeled, and an accompanying catalog provided the name by which the tree was known locally, its botanical name, a physical description, the size that its trunk grew to, the type of ground where it grew, the time of year when it flowered, and what it was often used for.[65]

The wood sample project was presented to the public in Spain's pavilion in 1851, subsuming this display of Cuban natural resources under the aegis of the crown and the empire. This wood sample project would have showed visitors to Spain's pavilion the natural abundance of Cuba and the possible commercial applications for these materials. The project also can be seen as elegiac considering the dawning awareness of the deforestation that was accompanying Cuba's rapidly increasing sugar production. It has been calculated that before the arrival of Columbus, the forests of Cuba had occupied more than 60 percent of the surface of the island but that by 1852, this had shrunk to 40 percent, and in 1923, to 16 percent.[66] Manuel Moreno Fragonals estimates that 13,420 hectares of virgin forest were being leveled annually in 1819, and 26,840 in 1830.[67] The learned and well-connected men of the Sociedad Patriótica would have been fully aware of this process, as naturalists had written about the dangers of deforestation in Cuba as early as 1792. Sagra himself would go on to

Figure 2.6. Edouard Willmann, G. B. Haase, and F. Chardon, *Panorama de la Habana,* 1855, engraving. (Library of Congress, Prints and Photographs Division, Washington, D.C.)

speak out frequently against the extreme deforestation that accompanied the expansion of sugar, most explicitly in an official report in 1860.[68]

Cuba's display at the Exposition Universelle des Beaux-Arts in Paris in 1855, by contrast to the 1851 wood sample presentation, consisted almost entirely of processed tobacco and sugar, displayed by individual companies and plantation owners.[69] Also included in the exhibition, though possibly not at Cuba's pavilion, was an ambitious panoramic view of the city of Havana (figure 2.6). This print was based on a daguerreotype by G. B. Haase and was printed in Paris. It shows a view of Havana from above, with palm trees and vegetation in the foreground. A road stretches away from the viewer, leading into the city. Several simple country houses stand nearest to the viewer on this road. It is only in the far background that the density of Havana can be seen—the palm trees and the rolling hills in the foreground are given significantly more visual emphasis than the buildings of the city itself. This print, then, represents Havana as a quaint backward city defined more by palm trees than by architecture.

A few Cuban individuals took part in the Exposition Universelle in Paris of 1867, including a guitar maker and a honey producer, but the island did not have its own specific pavilion. And in the United States Centennial Exposition of 1876 in Philadelphia, Cuba was again incorporated into Spain's display rather than given its own section, perhaps because of the escalation of the battle for independence on the island. As part of Spain's pavilion in the 1876 exposition, Cuban products on display included various manufactured chemicals, furniture, hats, embroidery, balsam, and a miniature display of Cuban railroad technology.[70] Thus, between 1855 and 1876, Cuba's presence at international expositions shifted from displays of raw agricultural products back to displays of technology and manufactured products.

Cuba did not participate in expositions for several years after the Philadelphia event and then had its own display at the Internationale Koloniale en Uitvoerhandel Tentoonstelling (International Colonial and Export Exhibition) in Amsterdam in 1883. At this time, the Ten Years' War (the first widespread military struggle for Cuban independence) had ended with the Pact of Zanjón, emancipation of enslaved persons across the island, and an uneasy peace between Cuba and the Spanish crown that still nominally controlled the island. Felipe Poey was one of the organizers of Cuba's pavilion at the Amsterdam exhibition, but little information exists about what Cuba included in its display, only that one of its central elements was a version of an illustrated ichthyology created by Poey—a multivolume publication that was to include all the known species of Cuban fishes.[71]

We know, then, that in the immediate aftermath of battles for independence, Cuba's Sociedad Patriótica spent significant sums of money and expended significant effort to send a set of products to an international exposition in Amsterdam. The central element of this display was a scientific text of Cuban fishes, produced by the man who had taken criollo Cuban citizen science as his directive throughout his life. At this fraught political moment, displays of Cuban biodiversity had again become salient and worth investing in. Just like at the 1847 agricultural exposition and the natural history museum, the display of Cuban nature was used at the Amsterdam exhibition to articulate a uniquely Cuban identity. Furthermore, although these previous spaces had been implicitly proto-nationalist, the Amsterdam display could make this claim for independence more explicitly in the wake of a Cuban revolution.

Overall, the various public displays became spaces for political imagination that were enhanced by reference to Cuban nature. El Templete and the new paseos of Havana worked in part to allow their white visitors to perform rightful and ancient ownership of Cuban land under the real or painted shade of the ceiba trees. Felipe Poey planned a natural history museum that seems to have been intended to create a criollo citizen science and an affective relationship with this science through its canoes and glass cases of specimens. The agricultural exhibition planned by the Sociedad became, in the wake of the Year of the Lash, a place to invoke small, independent, white-owned family farms dotting the Cuban landscape. And then at the end of the century, each of these visions of Cuban future—the white-owned family farm, the ordered biodiversity, and the naturalized political ownership—were packaged for international consumption and presented to international audiences at the massive imagination factories of the late nineteenth-century exposition complex.

Nature in each of these spaces became a way to claim Cuban, especially criollo, identity sometimes fully outside of Spain's aegis. Just like in the botanical garden, each of these spaces was likely also used by the Cuban elite members of the Sociedad to articulate white Cuban identity and to claim white knowledge and white leisure. This articulation of white knowledge of Cuban nature, in particular, would be further nuanced in the many printed projects that were produced in the years to come. In the next chapter, I will examine how many of these same tropes—the teeming biodiversity of Cuba, the white knowledge of this diversity—also drove the production of several ambitious natural history publishing projects. These books and prints were released to Cuban and international publics at the same time that the public spaces I've examined were being opened to these publics, and worked to reinforce many of the same political claims.

3. "The Novelty of the Productions of the Island of Cuba"

Natural History Illustration

As built spaces do, prints can also reveal the contested positioning of a group of people, and a print produced in 1787 by a sixteen-year-old boy presents us with a truly rich and strange tapestry of racialized scientific imagination. This is the frontispiece of *Descripción de Diferentes Piezas de Historia Natural las Mas del Ramo Marítimo,* also known as *Peces y Cruscáceos de la Isla de Cuba,* written by the Portuguese-born Antonio Parra and illustrated by his son, Manual Antonio Parra (figure 3.1).[1]

In this print, two figures coded as Native American—naked except for small loincloths, armbands, large earrings, and beaded necklaces, with heads shaved other than a small ponytail at the top—stand on an elaborately carved baroque table and hold up a net. The net is filled with dozens of different species of fish that manage to both overlap and tumble over each other and at the same time present a side of their body to the viewer for scientific identification. Above this dense net stands a small figure of Poseidon.

In the visual logic of this image, the rich natural wealth of the Cuban seas, and the diverse species of fish they contained, are being offered up joyfully to the reader of the volume. The two Native American figures—based, as we might have guessed, much more closely on previous depictions of an imagined generalized "Indian figure" as seen in books like Thomas De Bry's *Collectiones*

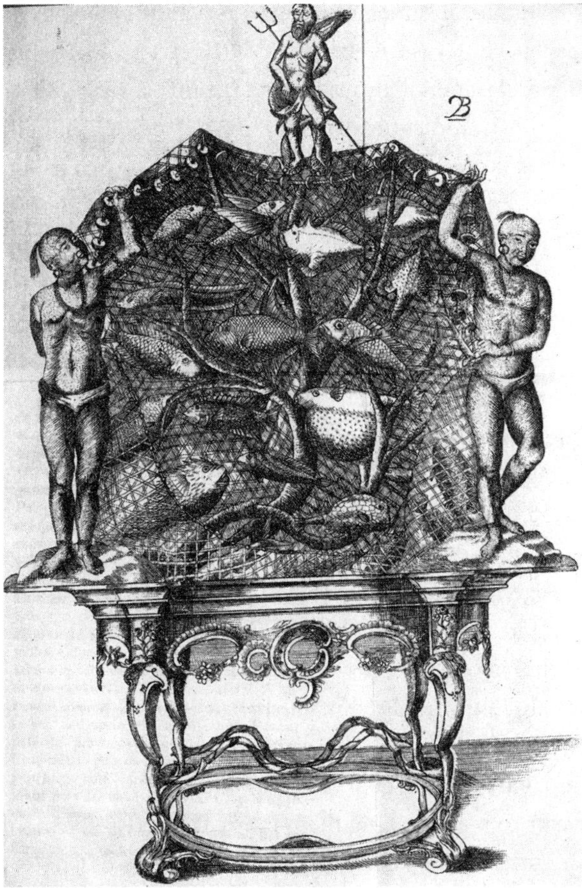

Figure 3.1. Manual Antonio Parra, frontispiece, 1787, engraving. (From Antonio Parra, *Descripción de Diferentes Piezas de Historia Natural las Mas del Ramo Marítimo* [Havana: Imprenta de la Capitanía General, 1787]; Courtesy of the John Carter Brown Library, Providence, Rhode Island)

Peregrinationum than on accurate depictions of indigenous Taíno costume—should be read as representing the island of Cuba itself, a personification of place.[2] Cuban nature is being offered to the viewer by Cuba itself, all presented on a table representing Europeanized good taste and cultural achievement. Above the tangle of rich biodiversity, a white-coded god stands, also smiling warmly at the reader. Through these layers of racialization, invitation, and visibility, the young Manual Antonio Parra in many ways anticipated the next one hundred years of natural history illustration on the island of Cuba.

Just as the botanical garden of Havana and museums and exhibitions had become physical spaces for the white elites of the city to enact shifting versions of white identity and imperial loyalty during the mid-nineteenth century, so too did the pages of illustrated natural history texts. Even as natural history publications explicitly concerned themselves with providing scientific information about the various indigenous species of Cuba to international audiences, the implicit aim of many of these publications was also to claim knowledge of Cuban nature for white scientists and to present a tempting vision of Cuban nature to potential white settlers, much like the botanical garden of Havana was meant to. The primary comparison that I will draw out is between illustrated books produced by the peninsular-born Ramón de la Sagra and the criollo Felipe Poey, but I also discuss several other important natural history texts of the same era. These include Juan Lembeye Lartaud, *Aves de la Isla de Cuba,* 1850, Francisco Sauvalle's planned illustrated *Flora Cubana,* circa 1875, and Felipe Poey's *Ictiología Cubana,* mentioned in the previous chapter, which he worked on until he died in 1891.[3] Together, this corpus of natural history illustration significantly expands and complicates the imaginary of Cuba's nature that I have outlined in the first half of this book.

First, though, I outline the short history of scientific publishing in Cuba. Due to the lack of printing technology on the island at the end of the eighteenth century, Antonio Parra's book and his son's illustrations remained somewhat crude and difficult to read, making them more of interest for their imaginative ideological content than their scientific representations of Cuban fish. The following years witnessed two ambitious hand-drawn compendia of Cuban natural history specimens, both tied to Spanish botanical expeditions discussed in Chapter 1 but neither published.[4] It was only with expanded print technologies on the island, especially the introduction of lithography to Cuba in 1822, that illustrated natural history publications truly began to flourish. Scholars including Juan Sanchez and Zoila Lapique Becali have studied the rapidly expanding universe of Cuban printmaking in the nineteenth century.[5] A single Cuban lithographic shop produced many different types of material, from maps to landscape prints to popular advertisements. It is partly because the same artists and printmakers worked across genres in this way that natural history illustrations from Cuba of the period show influence from these other genres.

Even though this Cuban printmaking industry was expanding at the time, Ramón de la Sagra's *Historia Física, Política y Natural de la Isla de Cuba,* published from 1838 to 1861, was written entirely in Paris in collaboration with professional scientists and scientific illustrators living there.[6] It argued implicitly and explicitly throughout its text and illustration program that Cuban nature was best understood from the vantage point of the European metropole by these professional white scientists. Felipe Poey's *Memorias Sobre la Historia Natural de la Isla de Cuba,* published by Cuban print shops in two volumes in 1851 and 1858, on the other hand, extended his citizen science program from his natural history museum to the printed page.[7] It argued that sustained, personal engagement with Cuban nature, especially engagement by criollos on the island, was necessary for knowledge of this nature. The two proposals put forth in these ambitious publications imagined different relationships between Cuba and imperial and extra-imperial networks. Sagra proposed that Cuba should align itself with European metropoles outside of the Spanish Imperial orbit, while Poey proposed a criollo Cuba that pushed away from imperial power by turning inward.

"The beautiful journeys": Ramón de la Sagra's Historia Física, Política y Natural de la Isla de Cuba

As the director of Havana's botanical garden, Ramón de la Sagra had redesigned the space into an agricultural experimentation station in which leading technology and cultivars from around the world were displayed. This display, in turn, was meant to drive white immigration to Cuba, in large part to counteract the growing political power of brown and black Cubans. Sagra's post-garden project, his multivolume *Historia Física, Política y Natural de la Isla de Cuba,* was also intended to drive white immigration to the island. However, the visual and textual rhetoric and the mode that Sagra used in service of this goal changed completely between the two projects. *Historia Física* celebrated the diversity of indigenous species, flora, and fauna found on the island of Cuba, a biodiversity that had remained at the margins of Sagra's version of the botanical garden. In producing *Historia Física,* Sagra collaborated closely with French natural history networks and referred extensively to previous publications by French and Prussian scientists. Sagra's Cuba, then, was extra-imperial.

By August 1833, in large part because of increasing interpersonal conflicts with other members of the Sociedad, Sagra had begun to make arrangements to leave Cuba, finally departing in 1835.[8] At first, he claimed to have intended to make a short visit to Spain and the United States, and to France for further botanical research and other sociological studies, after which he would return to Cuba. However, Sagra repeatedly requested that his travel license be extended. In March of 1835, he announced to his friend back in Cuba, Tomás Gener Bohigas, that he wanted to write a natural history of the island.[9] The result of this plan, *Historia Física, Política y Natural de la Isla de Cuba,* was published in Paris between 1838 and 1861 by the Paris-based publishers Maude et Renou and Bertrand in French and Spanish translations. *Historia* ultimately included 269 engraved illustrations, many of which are sumptuously hand-colored, and eleven volumes of text. In total, twelve different artists created the original drawings and paintings and thirty-nine different engravers worked on the illustrations. These include forty-two illustrations of birds and mammals, twenty-eight of reptiles, twenty-nine of mollusks, twelve of foraminifera (a type of nearly microscopic animal with a shell), twenty of insects, twenty of cryptogamae (mosses and fungi), and sixty-nine of plants. About a third of these illustrations include more than one species on the page together, often many species. The total number of Cuban species pictured across all of the volumes is well above 900. These numbers themselves speak to Sagra's extreme ambition for the project and his desire to create a total picture of Cuban nature.

Sagra's prospectus, published in 1837, also spoke quite clearly of his ambition for the project. Though this likely would have been written by the publisher instead of by Sagra himself, it is almost certain that Sagra was closely involved in its production. A close reading of this document reveals how the naturalist was conceiving of his project. The prospectus, written primarily for European audiences, traded heavily on tropes of the lush and generative tropics. It cited the fertility of the land, the "oxygenated breezes of the tropical region," and the "spontaneous vegetation of tropical plants." One particularly evocative sentence presented a vision of vegetation springing nearly spontaneously from the air in a tangle of life: "the vegetative life [in Cuba] is in such a degree of action and intensity that the surface of the earth is not enough to lend support to the immense mass of plants that invade the atmosphere to look for air, light and humidity, active agents of a lively, energetic and expansive existence."[10]

This language responded directly to European fascination with tropical regions at the time. As Mimi Sheller, alongside other scholars, has discussed, there was a renewed interest in tropes of "wild nature," embodied experiences with tropical nature, and the natural productivity of the tropics during the nineteenth century on the part of European and North American audiences.[11] These publications were fostered in part by a rise in international interest in the region in the wake of Alexander von Humboldt's publications. Beginning in 1807,

Humboldt wrote extensively on his travels to South America and other equatorial locations, his lush descriptions of natural splendor and vitality of the region creating an entire subgenre of tropical science and tropical travel narrative that authors from across the globe expanded on. This genre of tropical writing often focused on the tropics as sites of superabundance. Felix Driver and Luciana Martins describe Humboldtian science as visually driven, imagining the world not as a patchwork of disparate regions but as a spectrum of zones through which flora and fauna might circulate, and attentive to the observer's own entanglement and transformation.[12]

Humboldt also drove international interest in Cuba in particular with his *Political Essay on the Island of Cuba,* first published in 1825.[13] Sagra's prospectus referenced Humboldt repeatedly. Sagra's description, for instance, of a dense tropical rainforest echoed Humboldt's of the "luxuriant abundance of flowers and foliage . . . this exuberant and tangled web" of present-day Venezuela.[14] Sagra used this language precisely because it was a trope at the time of his writing, since he hoped to tap into this growing European fascination with the exotic tropics. This was despite the fact that many of these tropical forests had been cut down in Cuba to feed the growing sugar industry and even though the majority of the flora and fauna described in Sagra's *Historia* were from nonrainforest regions of the island.

While in Cuba, Sagra had repeatedly referred to himself as a true Cuban and wrote often about "our Cuba," and "our island."[15] This language changed significantly in his prospectus, however. The document began with an account of the "discovery" of America by "Columbus and his successors" (scientists including Étienne Geoffroy Saint-Hilaire, Prince Maximilian of Wied-Neuwied, Carlos Martius, Alcide d'Orbigny, and Aime Bonpland). This frame placed Sagra within this series of scientist explorers traveling from Europe to the tropical regions. Sagra reinforced this when describing his own arrival to the island—embarking from Spain in 1823 and following the path of Alexander von Humboldt, he "repeated the experiences and observations of [Humboldt] and other travelers." A Spanish governmental report on Sagra's decision to publish his project in Paris confirmed the parallels between Sagra's work and that of European naturalists, stating "that works similar to his are published in Paris under the auspices of the French government, such as M. d'Orbigny's journey, the expedition to the Mórea, History Natural of the Canary Islands etc."[16]

The discussion in the prospectus of the illustration program for *Historia* also referred to this legacy of traveler scientists. "As for its material execution, the plates have been entrusted to the skilled artists who currently perform similar ones for the beautiful journeys of the Coquille corvette around the world, of the lords Victor Jacquemont to India, d'Orbigny to the South America, Bory from St. Vincent to the Morea, Webb and Berthellot to the Canary Islands, whose publications the French Government protects." It was travelogues, then, that were named as models for the illustration program, especially travelogues to other tropical locations.

This is all to say that in writing a prospectus advertising a new project making Cuban nature visible for the first time, Sagra also reimagined how best to understand and represent this nature, especially when packaging it for European audiences. While serving as director of Havana's botanical garden he had focused on agricultural innovation and modernized small farms, but in Europe Sagra created a vision of Cuba as an untamed tropical paradise, teeming with wildlife. Furthermore, even though he frequently insisted that he was a "true Cuban" while on the island, in this prospectus Sagra positioned himself as a cosmopolitan European who had voyaged to the aforementioned exotic land. These two shifts in conception should be understood not just as strategic ways to drive consumer and subscriber interest in his new book, though they were certainly this too. They should also be understood as a reframing of Sagra's very racial and ontological argument about Cuban nature. Throughout *Historia,* he insisted that the unruly Cuban nature he described in the prospectus should be removed from the island and brought across the Atlantic to metropolitan Europe for study by experts there. For Sagra, rather than the Sociedad of Havana, for example, serving as the drivers of knowledge about Cuban nature, only professionals employed by European academic institutions were able to create this knowledge. It was a tacit denial of any possibility of criollo knowledge, of understanding Cuba from an embodied, embedded perspective. Further examination of Sagra's illustrations and text, and a comparison of his work with his primary rival, Felipe Poey, will tease out the implications of these claims at this moment of rising political tensions.

Of the 269 total engraved illustrations in Sagra's *Historia,* many of them hand-colored, I first examine the formal characteristics of those that picture only a single bird, mammal, or reptile, before examining the illustrations containing multiple species and finally those that picture a single plant. Each of the illustrations of a single species of an animal shows it surrounded by the white of the page—none include a background or setting of any kind. Furthermore, most are quite small relative to the page, enhancing the visual impact of the surrounding white space. A typical example of the illustrations in Sagra's *Historia* is the *Trogon temnurus,* in the Birds volume (plate 3). Drawn by the French draftsman Édouard Traviès, the bird is depicted from the side, perching on a branch. The visual emphasis is on the unique feathers and coloring of the bird, especially the fan shapes of the feathers of its tail. The bird is surrounded only by the white of the page—there is no background and no naturalistic shadows. It takes up only about a third of the page, further emphasizing the blank expanse around it. The branch on which the bird sits, rather than being depicted as any particular species of tree, remains a mere visual shorthand, depicted in loose sketchy lines, attenuated to its smallest possible size, and unidentifiable as any particular species of plant.

This approach to botanical illustration—a single specimen presented against a white background—was a fairly common format at the time, having been used for example in the eighteenth-century botanies of imperial colonies discussed by Beth Fowkes Tobin. As Tobin writes, "The white border and snipped twig of late-eighteenth-century botanical illustrations reinforced the idea that a plant could be plucked from one cultural and ecological context and inserted into another with ease and with little regard for negative consequences."[17] This format was seen more recently in one of the illustrated natural histories that Sagra referenced often throughout his text, Alcide d'Orbigny's popular account of his travels through Central and South America.[18] Sagra worked closely with d'Orbigny on the ornithology volume of *Historia Física* that *Trogon temnurus* is pictured in—D'Orbigny wrote the text of the volume and Sagra edited and oversaw illustrations.

This was not the only possible model available for depicting new species, however. The *Trogon temnurus* composition is notably different from other illustrations by the same artist, such as his *Picus viridis* from the 1857 edition of his set of lithographic prints, *Les Oiseaux les plus Remarquables* (figure 3.2).[19] In this print, the parrot is depicted perched on a palm leaf, about to eat a seed, with insects and butterflies nearby and royal palms in the distance. The parrot is thus surrounded by, and partially understood through, the other plants and animals of the place where he lives. In other words, it is an inextricable part of a local ecosystem. This "ecosystem" illustration mode was especially popular in the century before, in publications such as Maria Sibylla Merian's *Metamorphosis Insectorum Surinamensium* (1705) and Mark Catesby's *Natural History of Carolina, Florida, and the Bahamas* (1729–1747).[20] In both Merian's and Catesby's illustrations, the insects, birds, and animals they depicted are entwined, sometimes,

Figure 3.2. Édouard Traviès, "Picus viridis," 1857, hand-colored lithograph. (From Édouard Traviès, *Les Oiseaux les plus Remarquables par Leurs Formes et Leurs Couleurs* [Paris: Berrieux and E. Gambart, 1857]; The History Collection/Alamy Stock Photo)

literally, in the plants that grow in their habitat, their food and prey, and other animals that live nearby. For Sagra's *Trogon temnurus,* and likely in response to Sagra's direct request, Traviès removed all of these elements.

This minor detail changes the meaning of the illustration significantly. By removing this environment, the bird illustration produced for Sagra's project makes a visual argument that this animal can be extracted from its world and moved through space. Furthermore, it makes a visual argument that an animal does not require its accompanying ecosystem to be understood, and in fact is better apprehended without its surroundings. All of the illustrations in Sagra's *Historia* show specimens similarly removed from environments and floating against white backgrounds. Most of them being quite small on the page reinforces this sense of removal and mobility. These illustrations support the claims in the text that only white professionals in Europe could understand Cuban nature. When he oversaw the production of his *Trogon temnurus* illustration, Sagra had been reading and collecting natural history publications and visiting scientific libraries and collections for decades. He would have been extremely familiar with the range of illustrated natural histories that were being produced in Paris, London, and elsewhere and seems to have deliberately chosen the style of illustration that reinforced important themes in his text.

Since *Historia Física* was itself written in Paris, using specimens that Sagra had collected during his time in Cuba, many of the artists employed by Sagra and his publisher had worked on projects at the Muséum d'Histoire Naturelle or the Jardín des Plantes in the city, while others had worked on previous natural history books released by Sagra's publisher. In addition to the French illustrators mentioned, each of the eleven volumes was primarily written by a French naturalist under Sagra's supervision.[21] In this way Sagra chose professional European scientists to write about Cuban nature and professional European artists to depict it—another tacit claim that only (white) European professionals could understand Cuban nature.

A close examination of Sagra's text further shows his insistence in this project on meaning-making in Europe through the transatlantic transportation of specimens. This belief can be read, for instance, in Sagra's discussion of the *Crocodilus rhombifer,* or caiman, from the volume on reptiles, as well as its accompanying illustration (plate 4). The illustration of the caiman pictures the animal as if the artist is standing to its side and slightly above it. The animal's legs are slightly splayed, its tail curves down toward the text below and its head turns up toward the corner of the page. The central image of the caiman is accompanied by other illustrations in grayscale showing anatomical details such as the underside of the animal's snout, the anatomy of its foot, and a cross-section of its tail. The caiman and the anatomical details that surround it, which were engraved by Gabriel (full name unknown) after a drawing by Jean-Gabriel Prêtre, float against an expanse of white. As in the bird illustration discussed above, there is no hint of a background behind the caiman or shading beneath its feet. These feet even point in different directions and the animal is placed at the top of the page, almost as if it has floated upward and is flailing in the air. This is an illustration that serves to communicate anatomical details about a new Cuban species but also serves to give the viewer a sense of this species untethered from its environment and moving through space.

The description of this species in the text that accompanied the illustration also charted a process of extraction and movement along with a final reconsolidation in France. According to this text, Georges Cuvier found a skull of a reptile now known to be this caiman in the storage of the Muséum d'Histoire Naturelle in Paris and published it in 1825, though he still didn't know the origin of the animal. A Mr. Wiegmann discovered a similar skull in 1828 in a collection brought from Mexico, but his note on this discovery "had passed unnoticed" until Sagra sent another skull and finally a complete specimen from Cuba.[22]

This is a story of an encounter in the European metropole with an object that had been removed from its original local context. Without this context, only certain types of information could have been gleaned from the object, such as the shape of its skull. As specimens and text moved from tropical ecosystems into European museum spaces, the little information that did exist was often lost or corrupted—as in the "unnoticed" note from Mr. Wiegmann. Skulls and notes in this account did not prove sufficient to convey the most crucial information about a specimen. It was only when Sagra shipped a young complete specimen of the caiman to the Museum of Paris in 1833 and when this specimen was put in conversation in Paris with the other parts of specimens that the animal was able to be fully identified and its habitat and range known. In the logic of this anecdote, the Museum of Paris serves as the necessary site for the production of knowledge—not, as we might imagine, the original environment of the caiman. Sagra makes no mention of observing the living animal in the wild, for instance. Embodied observational data is not of use in this model of knowledge production, only trained (white) professionals housed in a European academic setting.

Other volumes in *Historia* discussed similar processes of extracting specimens from Cuba and studying them in Europe. The introduction of the volume on Ornithology, for example, included Alcide d'Orbigny's statement that "Mr. de la Sagra gave us, along with his ornithological, conchilogical, and other collections, his notes and those on the birds of Cuba made by Mr. Poey, distinguished naturalist of the island. *To complete them, we have added our personal observations.*"[23] Throughout the volume, d'Orbigny frequently discussed this process of combining, in Paris, various pieces of information about the flora and fauna he described. His description of the pine falcon cited observations by William Jardín, Alexander Wilson, and Thomas Hutchins.[24] A similar process was discussed by Camille Montagne in the volume on mosses and fungi: "A sample of this species, from Brazil, was first given to me without a name by Persoon & Bonnemaison. . . . I received from MM. Bélanger and Lehmann the same alga harvested at the Cape of Good Hope. Mr. Maille gave it to me from Haiti, and Mr. Webb collected it himself from an islet or rather a rock called Alboran, located in the strait of Gibraltar."[25]

This style of writing, with its layers of citations of other scientists, was not uncommon at the time but is noteworthy nonetheless. In making this decision, and in tracing the knowledge production networks of the objects in *Historia Física, Política y Natural de la Isla de Cuba* in this particular mode, Sagra and his

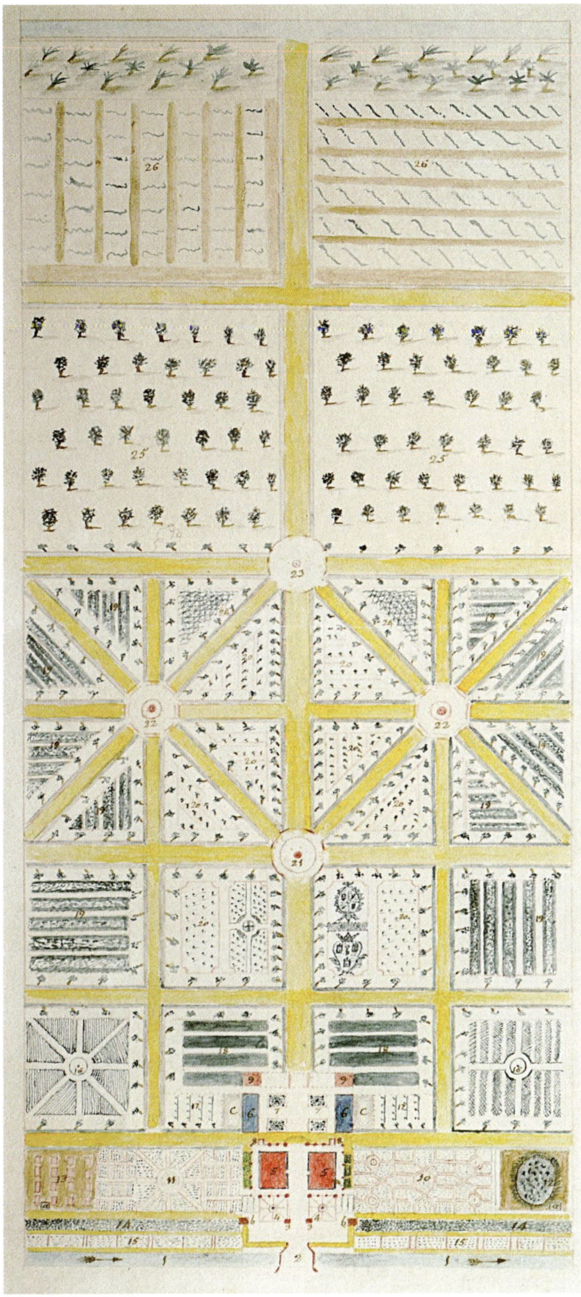

Plate 1. Francisco Lemaur, *Plan y Distribución del Jardín de la Habana,* 1821, ink and watercolor. (Archivo General de Indias, Seville)

Plate 2. Jean-Baptiste Vermay, *The First Cabildo,* circa 1827–1828, oil on canvas. (Oficina del Historiador de la Ciudad, Havana)

Plate 3. Guyard (probably Jean-Baptiste Guyard), after Édouard Traviès, "Trogon temnurus," circa 1845, hand-colored engraving. (From Ramón de la Sagra and Alcide d'Orbigny, *Historia Física, Política y Natural de la Isla de Cuba, Atlas Aves* [Paris: Maude et Renou and Bertrand, 1838–1861]; New York Public Library)

Plate 4. Gabriel (possibly Charles-Louis Gabriel), after Jean-Gabriel Prêtre, "Crocodilus rhombifer," 1838–1861, hand-colored engraving. (From Ramón de la Sagra, Jean Théodore Cocteau, and Gabriel Bibron, *Historia Física, Política y Natural de la Isla de Cuba, Atlas Reptila* [Paris: Maude et Renou and Bertrand, 1838–1861]; New York Public Library)

Plate 5. Frédéric Mialhe, "Almiquí," 1851, lithograph. (From Felipe Poey, *Memorias Sobre la Historia Natural de la Isla de Cuba,* volume 1 [Havana: Impr. de Barcina, 1851]; Beinecke Rare Book and Manuscript Library, Yale University)

Plate 6. Probably Francisco Sauvalle, *Metastomacea,* circa 1873–1879, watercolor and ink. (The LuEsther T. Mertz Library of the New York Botanical Garden)

Plate 7. Frédéric Mialhe, "Fenómeno Natural en el Potrero de Dn. Manuel Lomba from Moreau," 1839–1840, lithograph. (From *La Isla de Cuba Pintoresca* [Havana: Litografía de la Real Sociedad Patriótica, 1839]; New-York Historical Society)

Plate 8. Eduardo Laplante, "Ingenio Flor de Cuba," 1855–1857, lithograph. (From Justo Cantero, *Los Ingenios* [Havana: Impreso en la litografía de Luis Marquier, 1855–1857]; Beinecke Rare Book and Manuscript Library, Yale University)

Plate 9. Eduardo Laplante, "Ingenio La Ponina," 1855–1857, lithograph. (From Justo Cantero, *Los Ingenios* [Havana: Impreso en la litografía de Luis Marquier, 1855–1857]; Beinecke Rare Book and Manuscript Library, Yale University)

Plate 10. Marquilla cigarrera (illustrated cigarette packaging) printed by the company Dignidad Gran Fábrica, 1860–1870, chromolithograph. (Biblioteca Nacional de España, Madrid)

collaborators insisted that science could only have been produced by professionalized experts in the European metropole despite the losses and corruption that often accompanied this process. Again, this was the process by which *Historia* itself was produced—Sagra extracted the materials for his history from Cuba but then refined them (had them drawn, printed, and described) in Europe.[26] Of course, this was not a unique decision on Sagra's part. The reason that he chose Paris as his center of production was because publishers in the city were printing, at this same time, illustrated natural histories about locations from around the world. However, it was still a deliberate choice that reveals a political and geographic orientation on Sagra's part away from Cuban scientific knowledge production on the island itself.

Sagra's *Historia Física,* created at a time of rising racial unrest on the island of Cuba, had a racializing dimension as well. First, as described, the very decision to locate the production of this volume in professional scientific circles in Europe can be understood as a claim that educated white European elites were better situated to understand the natural history of the island of Cuba than Cubans themselves, including black or brown Cubans. Sagra also seems to have hoped that his volumes would continue to drive white immigrants to the island, just like his version of the island was meant to do. Sagra repeatedly discussed this white immigration in the text of *Historia.* In volume 2, for example, he wrote that "a system of agriculture and rural economy based on scientific principles will be absolutely impossible to organize in [Cuba] as long as the cultivation of sugar is done by slave labor, and that it will only be when it is done by free men that it will be able to acquire a happy and rapid growth." He then went on to restate the necessity of replacing enslaved workers with "white settlers, active, honest, hardworking and able to withstand the climate."[27]

For Sagra, the negative environmental impacts of plantation sugar—he described how this monoculture depletes the soil—and the negative social impacts of slavery, which he argued "paralyzes both the natural forces of the soil and the intellectual means of man," were intertwined.[28] The solution to both, he believed, as he had imagined in his early life at the botanical garden of Havana, was diversified farming done by white immigrants. Unlike at the botanical garden, however, Sagra did not focus, in *Historia,* on the productive potential of imported cash crops to encourage this immigration. Instead, he seems to have imagined that the biodiversity and fecundity that was showcased in *Historia* itself would be the driving force for this immigration.

The very production process of the volumes may have served to emphasize this biodiversity, with the publication covering so many different types of Cuban flora and fauna. The 269 illustrations of *Historia* would have been released in sets as they were printed and then collated in the massive atlas volumes held by library collections today. The release of these sets of illustrations from the printer did not seem to have followed any clear order—a set of prints of birds might be released one month and then a set of plants another, for instance.[29] While most illustrated natural history volumes (and other sets of prints) were produced in this sequential manner at this time, the scale and diversity of objects represented in Sagra's illustrations, the long duration of the production process, and the extreme lack of order in releasing sets would, I believe, have added to the reader's sense of the nearly infinite biodiversity of the island of Cuba. This is the same biodiversity that Sagra discussed in the prospectus of his project as teeming with "action and intensity."

Despite, or perhaps even because of, this somewhat disorganized production process, many of the illustrations in Sagra's *Historia* emphasize pattern and visual order. This order is most apparent in the illustrations of *Historia* that feature multiple species arranged on the page together. An example of these is plate 23 of the Mollusca volume, a compendium of the shells of seventeen species of mollusks designed by Delarue (likely Jean de la Rue) and engraved by L. Rebuffet (figure 3.3). The shells are arranged within a grid, neatly lined up along the center of the page and across its edges. The largest shells are placed at the center of the composition, with smaller shells radiating outward. The overall impression of the illustration is of a pattern organized according to symmetry and increasing size, but with visual interest added by breaks to this symmetry. The composition becomes both a decorative pattern and a grid of specimens that can be compared.

This gridlike arrangement of multiple species on a page was a common visual technique at this time. The closest reference for Sagra's grids seems also to have been from Alcide d'Orbigny's *La Relation du Voyage,* as in the first illustration from d'Orbigny's volume on insects (figure 3.4). For Sagra, these grids likely served to fix this natural fecundity of Cuba into orderly rows and columns for the visual delectation of potential white immigrants. These grids became emblems of Sagra's vision of the professionalized scientific knowledge that the nature of Cuba was processed into in the European metropole.

It is also possible that the patterns and grids that structured many of Sagra's illustrations were related to white elite interest in Cuba at the same time

Figure 3.3. L. Rebuffet, after Delarue (possibly Jean-Baptiste Antoine de la Rue), untitled, 1838–1861, hand-colored engraving. (From Ramón de la Sagra and Alcide d'Orbigny, *Historia Física, Política y Natural de la Isla de Cuba, Atlas Mollusca* [Paris: Maude et Renou and Bertrand, 1838–1861]; New York Public Library)

in defining and reinforcing systems of social classification, including race. Cuban enslavers, for instance, were extremely interested in the ethnic origin of the people they enslaved and developed complex theories about the temperament and physical attributes thought to be possessed by people that came from different regions of Africa.[30] Sagra fully adhered to contemporary theories of racial difference. In 1828, for example, Sagra wrote about black people being a distinct race, with a simple "temperament," and free people of African descent being "vegetative in the most miserable state."[31] Sagra's illustrations, though not of human

Figure 3.4. Alphonse François, after Jean-Baptiste Antoine de la Rue, untitled, 1835–1847, hand-colored engraving. (From Charles Dessalines d'Orbigny, *Voyage dans l'Amérique Méridionale, à l'Intérieur de la Côteferme, et aux Îles de Cuba et de la Jamaïque* [Paris: Agen, Impr. de P. Noubel, 1835–1847]; Smithsonian Libraries)

beings themselves, implicitly reinforced the contemporaneous preoccupations with ordering, census-taking, and classification as an important step to addressing perceived population disparities.

Illustrations of vascular plants constitute almost one-third of the prints included in *Historia* and differ from the other illustrations discussed in some important respects. The majority of these illustrations were originally drawn by Antoine-Charles Vauthier, a French artist who was also known for his collections of plants and insects from Brazil.[32] While the other illustrations in *Historia* are in full color, with most of the page left as white background, the vascular plants were depicted in black and white, nearly filling the page, always with anatomical details below. These plant illustrations make visual arguments for a slightly different form of Eurocentric knowledge production than the other illustrations in *Historia* do. They picture Linnaean botany, previously consigned to the corner of Havana's botanical garden, now moved to the center of the page.

A typical example of these vascular plant illustrations is the *Clematis havanensis,* the first illustration in the Vascular Plants volume (figure 3.5). The

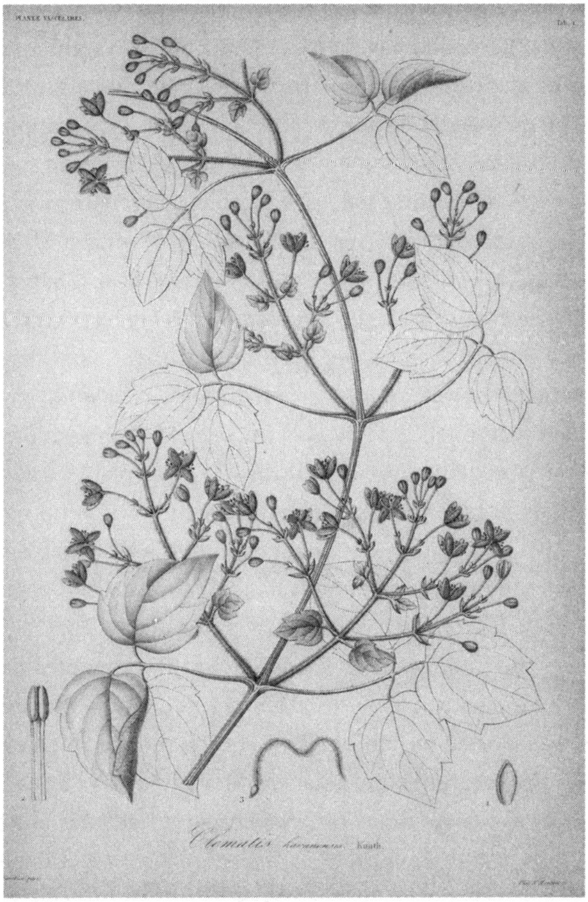

Figure 3.5. Plée & Jean Joseph Mougeot after Vaulthier, "Clematis havanensis," engraving. (From Ramón de la Sagra and J. F. C. Montagne, *Historia Física, Política y Natural de la Isla de Cuba, Atlas Plantas Celulares* [Paris: Maude et Renou and Bertrand, 1838–1861]; New York Public Library)

plant is arranged across the page so that its stem stretches in a long curve and its leaves press up against the picture plane for ease of viewing. Views of the pistil, stamen, and seed (details that are necessary for Linnaean typology) are arranged in an orderly line along the bottom of the page. The flowers of the clematis have been depicted in various stages of opening so that the life cycle of the plant can be seen by the reader all at once.

This print thus closely follows botanical illustration conventions of the period and appears designed primarily for a Linnaean assessment of this Cuban plant. There were other models available to Sagra and to Vauthier for these fifty-two engravings of plants, just as there had been other models for Sagra's fauna illustrations. The men could have used full color and tight cropping to fill the page with naturalistically twisting vines and flowers that reach toward the picture plane. This was the mode used, for example, in François Richard Tussac's *Flora Antillarum,* published between 1808 and 1827; and in Michel Descourtilz's *Flora Medicale et Pittoresque des Antilles,* 1821–1829.[33] Sagra would have been particularly aware of this style of illustration as he had worked on a translation of the Descourtilz book during his time in Cuba.[34] Sagra also could have chosen a format that showed an interwoven ecological network, with plants depicted alongside the animals that lived with them, as discussed previously.

Instead, in commissioning the illustrations of plants used in *Historia Física,* Sagra seems to have requested for the species of flora to be depicted in stark black and white, without any insects, animals, or other plants included, flattened against the picture plane and at the center of the page. The white spaces around these flattened leaves, flowers, and stems of Sagra's flora illustrations do some of the same work as the grids and the expanses of white page discussed earlier. They show the pictured species as extractable from the Cuban ecosystem and transported to Europe. They also fix the specimen onto the page, creating a sense of visual order and total knowledge. Sagra's *Clematis havanensis* represents in this way the epitome of the "reasoned image" that Lorraine Daston and Peter Gallison link to Linneaeus's *Hortus Cliffortianus.*[35] It has been rendered as an archetype of a plant that removes any details specific to or evoking an individual specimen. It is an idea of a *Clematis havanensis,* rather than one single and fallible version of the plant. And, like the other objects pictured in Sagra's *Historia,* it has also been stripped of any reference to ecosystem or habitat and reconstituted by European scientific professionals.

Despite the way the blank white backgrounds and stark borders of the pages of his illustrations hint at seamless extraction and circulation, the making of Sagra's *Historia* was a highly fraught and contentious process as was hinted at above. To carry out this project, he had to ship all of the specimens that he had collected during his years as director of the botanical garden across the Atlantic. Some of these cases were lost and others arrived with significant delay and decay. Sagra also faced a continuous lack of funds, including the money necessary to pay his illustrators and collaborators. Sagra's file at the Archivo General de Indias in Seville is filled with many dozens of letters begging the

Spanish crown for more money. A letter written on February 16, 1836, for instance, mentioned Sagra's challenges in paying for the transportation of the boxes of specimens that he collected in Cuba as well as his difficulties in paying "los artistas," presumably some of the artists whom he had already employed to start working on his *Historia.* He requested more funds to cover the handling of the boxes to expedite the process in the future.[36] Further delays were caused by, according to Sagra, inadequate funding and delays in payments of his salary.

In large part because of these delays, the archive also contains many complaints from the Spanish government about the publication schedule of *Historia.* One such letter, for instance, stated that Sagra had only published thirty sections of the book so far and that these sections were "disconnected from each other for the most part."[37] The Spanish government issued at least seven royal orders for Sagra to complete the publication and send the volumes to subscribers (in 1839, 1843, 1844, 1847, 1852, 1859, and 1863).[38] Spanish officials even produced several detailed charts of all the sections of *Historia* that had been received at various points in time and those that were still missing. At one point the demands from the Spanish government as well as from other subscribers had become so insistent that Sagra had a short pamphlet printed, titled "Recent documentation explaining the delays experienced in the publication of the work *Historia* . . . that is being produced by Ramón de la Sagra in Paris."[39] In the end, someone had to be commissioned by the Spanish government to go to Sagra's bookbinder and collect the outstanding copies of the publication themselves.[40] Sagra even began the project by acknowledging these challenges. He wrote in the introduction to volume 1 of *Historia* that "the writing of natural history, executed by a great number of collaborators and demanding considerable expenditures for the plates which must accompany it, made us feel and makes us still experience delays it is not possible for us to avoid."[41] Given how intensely he would have felt the pitfalls and losses that accompanied the transatlantic transportation of objects and knowledge, it is particularly noteworthy that Sagra's text and images repeatedly insist on the value of this transportation.

As might be expected due to the delays, Sagra's relationship with the Spanish crown became ever more strained over the course of the publishing of *Historia.* While the prospectus claimed that he had the support of Maria Christina, queen regent of Spain (1833–1840), and another paragraph of the prospectus mentioned the "political regeneration of Spain" as one of the goals of the project, Sagra seems not to have officially asked to dedicate *Historia* to the Spanish crown until 1836. An official note on the letter that contained his offer of dedication counseled that he should not be allowed to do so.[42] Perhaps because of this, none of the volumes of *Historia* bear a dedication on their title pages. This fact alone should remind us of the fragility of Madrid's reach—that while natural history in the Spanish Americas had been driven entirely by the crown at the end of the eighteenth century, by the time Sagra was publishing his volumes, he was doing so almost entirely outside the aegis of, and often in opposition to, the Spanish imperial administration.[43] I believe he chose to face these many obstacles because of his commitment to aligning the project and himself with professionalized European traveler naturalists, rather than with the local criollo science of Cuba.

While Sagra was struggling to transport his specimens across the ocean and was stretching to find funding to have his volumes printed in Paris, there were several publishers in Havana who were producing natural history publications and complex sets of prints. It would have been easier in many ways for Sagra to have used one of these publishers for his project. His fellow Sociedad member Felipe Poey deliberately chose to publish a competing illustrated book of Cuban natural history with some of these local publishers in Havana during the same years that Sagra was producing *Historia Física.* Poey discussed at length his decision to do so, one of his main reasons being the "advantage that his residence on the island gives him."[44] Like his museum had been intended to, Poey's book aligned squarely with criollo science, the same criollo science that Sagra rejected emphatically and at great cost.

"The advantage that my residence on the island gives me": Felipe Poey's Memorias Sobre la Historia Natural de la Isla de Cuba

In many ways, Felipe Poey's *Memorias Sobre la Historia Natural de la Isla de Cuba,* published with Francisco Louis Marquier's Litografía Cubana press between 1851 and 1858, was the direct counterpoint to Ramón de la Sagra's *Historia Física.* Poey's book was published entirely in Cuba, entirely by a criollo natural historian and using entirely local informants and observations, and with an entirely different mode of illustration from Sagra's.

Both volumes of Poey's *Memorias Sobre la Historia Natural* are structured as collections of articles with topics ranging from examinations of a specific species to indices listing many different species in relation to each other to Poey's writings on the relationship between color and divinity. Several of the articles had been first published elsewhere, either in Poey's *Centurie de Lépidop-*

tères de l'Ile de Cuba, first published in 1832, or in journals such as *Anales of the Lyceum of New York* and *Floresta Cubana.*

Together, the volumes contain fifty-three total illustrations, almost all of which were drawn by Poey himself. Poey was a self-trained draftsman who was especially interested in the comparative osteography pioneered by Henri-Marie Ducrotay de Blainville.[45] Blainville's comparative anatomy was a systematized way of depicting the skeletons of animals to facilitate visual comparison. Poey was deeply interested from the beginning of his career in how scientific illustrations could drive knowledge production. Furthermore, Poey wrote extensively throughout the text of both volumes of *Memorias Sobre la Historia Natural* about the illustrations he had designed and chosen to accompany the text. He discussed how to arrange specimens for illustration—for instance, the best position of mollusks on the page or his decision to represent fish always from one side, "for clarity."[46] Poey also discussed the lithography process itself and recent improvements to the process in chapter 11.[47] Poey did work with a professional artist, Pierre Toussaint Frédéric Mialhe, for a few of the illustrations on *Memorias Sobre la Historia Natural.* Even this relationship, though, was based on intimate, shared embodied experience of the Cuban landscape. Though born in France, Mialhe had lived in Cuba since 1838. He and Poey often traveled together on specimen-collecting and sketching journeys.

This all indicates that the illustrations in Poey's *Memorias Sobre la Historia Natural* should be considered an integral part of how the two volumes functioned, just as Sagra's illustrations had reinforced the claims of his text. They were carefully considered to produce specific affective and intellectual responses. It also means, furthermore, that Poey was thinking differently about illustrations in *Memorias Sobre la Historia Natural* than Sagra was in *Historia Física, Política y Natural de la Isla de Cuba.* Whereas Sagra commissioned illustrations from French expert natural history illustrators, Poey insisted on creating almost all of his drawings himself or working with an artist whom he knew to be intimately familiar with Cuban nature. The illustrations that Poey chose for his *Memorias* make an entirely different set of visual claims than those made by the illustrations in Sagra's *Historia.* Where Sagra's illustrations prioritized white space, objective remove, and orderly arrangement of information, Poey's illustrations are visually dense, often emotional and affective, and refuse order. These qualities served to reinforce his contention that Cuban nature should best be understood from Cuba itself.

Although both Poey and Sagra received their training in Europe and both men remained closely connected to European natural history networks throughout their careers, the Cuba-born Poey, throughout his book, deliberately contrasted his lifetime of experience with Cuban nature with Sagra's mere twelve years on the island. Even more importantly, Poey went on to insist throughout the book that intimate, embodied, long-term experience with Cuban nature (the kind that Sagra had had little of, and had been removed from for a long time when writing) was necessary to produce scientific knowledge. Related to this, Poey maintained a deep interest throughout his life in the natural historical knowledge held by non-elite members of Cuban society, who had also been at the center of his museum. Poey discussed his belief in the power of deep local knowledge throughout the text of *Memorias Sobre la Historia Natural,* again often by contrasting his project explicitly with Sagra's *Historia Física, Política y Natural de la Isla de Cuba.* There seems to have been at least some overlap between the intended audiences of this and of Sagra's *Historia*—the two were listed together in catalogs of science publications for sale in Edinburgh and are cited together in Spanish-language articles on mammals.[48]

In the introduction to the *Memorias Sobre la Historia Natural,* Poey wrote:

> Two main considerations have led me to publish these memoirs: 1, the novelty of the productions of the Island of Cuba, which in the animal kingdom are mostly exclusive to its soil; 2, the advantage that my residence on the Island gives me, as to represent with its natural colors the objects that are altered by death, by preservation in alcohol, and by the shipment of furs or herbs.[49]

This claim of Poey's advantage should be considered a direct challenge to Sagra's claims that natural history was best practiced in the European metropole. Poey reinforced his claims by referencing the losses that accompanied the preparation of specimens for shipment—the alteration of specimens through death, desiccation, and preservation. A contemporaneous discussion of Poey's project in the Paris-based *Revue Zoologique* concurred that Poey often had superior access to Cuban specimens without losses, stating that his work complemented Sagra's, "sometimes because of the species which M. de la Sagra could not procure, sometimes because of new facts."[50]

Throughout both volumes of *Memorias Sobre la Historia Natural,* Poey often made note of which "advantage" his "residence on the island" gave him. As he wrote about the almiquí, a small native Cuban rodent, "to the observations of Brandt, Jacome, Figueredo, Zenea and my own, those that are inferred from the [social] organization of the animal can be added; since nature offers in its unchanging aspect a book that does not lie and one discovers that in [nature's] mute language the dynamic state of beings and physiological phenomena are known."[51] Poey is referring quite explicitly to the kind of information that was lost in the process that Sagra followed in his publication—the ecological relationships among species, the knowledge gained from observing living animals, even the loss of color and texture suffered by skins and feathers as they made their way across the Atlantic to European naturalists like Sagra.

Poey also took care to describe his own collecting practices and his own encounters with the specimens he was describing. For example: "a few days ago I went to look for some fresh specimens in the Zanja of Havana to complete this article, and I found another species"; and "I have collected many individuals in the company of Mr. José Blain y Cervantes, a neighbor of Santa-Cruz."[52] At one point he even noted the flavor of a particular species of fish.[53] This type of information about the embodied encounter was rarely included in Sagra's *Historia.* Poey also included, whenever possible, details about exactly where on the island a particular species had been collected, even if it was not he who collected it.[54]

This text, then, proposed an alternative geography of Cuban science to that of Sagra's. Where Sagra argued that Cuban nature should be extracted from the island and brought to Europe to be understood there, Poey insisted, as he had in creating the museum of natural history, on Cuban science that was deeply situated on the island itself and that was driven by a long-standing personal connection to the terrain. Poey did cite European naturalists throughout the text, and also wrote about sending Cuban specimens for further study to, for instance, Louis Agassiz at Harvard and the Real Jardín Botánico of Madrid.[55] While Sagra claimed that knowledge was only produced in the European metropolis, however, Poey argued that European naturalists must come to Cuba and immerse themselves in the landscape there to truly understand Cuban nature. Chapter 49 in volume 2 of *Memorias Sobre la Historia Natural,* for example, is a long account of the German-born scientist Juan Gundlach's visit to the Matanzas region and his travels around this region in search of indigenous species of mollusks.[56] His account of collecting mollusks included descriptions of entering a cave near Guajaiban, the views from the hills of los Organos, and his travel up a mountain by foot. Gundlach had moved to Cuba in 1839, some fifteen years before the events he described.

This type of detailed travelogue focusing on the embodied experiences of the naturalist who was writing it was fairly common in the mid-nineteenth century. One of the best-known examples is Joseph Dalton Hooker's *Himalayan Journals,* first published in London in 1854.[57] However, such travelogues had not been included in any of the previous accounts of Cuban nature and were certainly not included in Sagra's *Historia.* Sagra's account instead contained only depersonalized descriptions of individual species, all within the frame of an overview of Cuban history, society, and economy that drew closely from Humboldt's previous account of Cuba. I believe that the inclusion of this dramatically narrated travel account in Poey's book was meant to argue that such intense embodied encounters with Cuban ecosystems were necessary to claim knowledge of Cuban nature.

Recall that Poey was writing *Memorias Sobre la Historia Natural* in the wake of his failure to establish a museum of natural history in Havana. The *Memorias de la Sociedad* of 1849 reported that Poey had agreed to surrender all of the collections of the museum to the University of Seville.[58] Another document from the same year discussed Poey's own dismissal and contained a list of all of the furniture that had been produced for the museum that was to be sold to recuperate costs.[59] Poey had worked on this museum for nearly a decade and had been continually thwarted by debates over its location and by funding challenges. To be forced to send this collection to peninsular Spain and to give up his dream of having a museum of natural history in Havana would have been quite a blow to Poey. This would have been particularly true because the museum, like his *Memorias Sobre la Historia Natural,* had also proposed a uniquely Cuban citizen science program. Given that he began compiling his *Memorias Sobre la Historia Natural* at almost the same time that his museum was being disbanded, it makes sense that Poey would have used the text of this publication to reiterate his calls for an embodied direct experience with Cuban nature.

It was partly this deep interest in long-term, embodied knowledge of Cuban nature that also led Poey to repeatedly celebrate throughout *Memorias Sobre la Historia Natural* the knowledge held by non-elite Cubans. For instance, Poey maintained a particular interest throughout his life in ichthyology, but rather than just study publications on the subject, he visited the fish markets of Havana to buy new specimens and to discuss fish anatomy with the fishermen there. An article written about Poey near the end of his life states that "every fisherman in Havana knows 'Don Felipe' and looks upon him as a personal friend."[60] In the planned introduction for his *Ictiología Cubana,* never published during his lifetime, Poey insists that "all the fish presented here have been acquired in the port of Havana, usually the fish are from the fishermen of Marianao, Cojímar, and Bacuranao."[61]

Another example of Poey's citation of the non-elite informants and other participants in the production of *Memorias Sobre la Historia Natural* is from his chapter on the almiquí. In this account, Poey described how a beekeeper had found the skull of the rodent in the mountains. The beekeeper gave the skull to a man he knew, who in turn gave it to his brother who was a priest in Guanabacoa. The priest owned the skull for three years, then gave it to Poey, who sent it to a scientist in London. Later, additional living animals were caught by an amateur naturalist from Bayamo, and one was killed and its pelt sent to the Royal University in Havana. There it was drawn by Mialhe.[62]

In this account, the original scientific knowledge that animated this movement was that of the beekeeper. This rural laborer had discovered the skull during his work in the mountains where he was closely connected to the cycles of local nature. His embodied knowledge combined with the textual knowledge of European scientific establishments to enable an understanding of the animal. In this account, then, Poey insisted that scientific study needed to be done in close collaboration with local informants and as close as possible to the site of origin. This use of non-native informants was not unusual for scientific publications of the mid-nineteenth century, but explicitly celebrating the ties between the scientific production of knowledge and the fish market was an unusual choice. Again, this is quite different from how Sagra described knowledge production in his *Historia.* For Sagra, scientific knowledge about Cuba was produced only when skulls and skeletons of animals were transported, discovered, and studied in a professionalized European setting as in, for instance, his discussion of the study of the caiman skull in the storage rooms of the natural history museum of Paris. For Poey, knowledge was, instead, produced in Cuba itself, by laborers and other non-elite criollo Cubans, not through professionalized training but through lifelong embodied knowledge of Cuban nature.

In fact, throughout *Memorias Sobre la Historia Natural,* Poey cited Cuban collectors and amateur naturalists. These include Cecilio Jácome, the nephew of the priest from Guanabacoa mentioned earlier; José María Velazquez, a practicing medical doctor who collected several species of mollusks for the book; and the engineer Julio Sagebien.[63] Unlike the professionally trained scientists that Ramón de la Sagra collaborated with to produce *Historia Física,* Poey produced his own version of Cuban natural history in collaboration with amateur naturalists. These individuals that Poey cited held expertise not because of their training or because of their academic affiliations, but because of their lifelong residence in Cuba and their personal interest in its flora and fauna. In some real respects, Poey's *Memorias Sobre la Historia Natural* was a collaborative image of Cuban nature produced by criollo Cubans themselves. Poey's *Memorias Sobre la Historia Natural* represents a kind of citizen science.

Poey claimed that fostering this knowledge, as well as an interest in science, was one of his primary goals in writing *Memorias.* "My work," he wrote, "addresses not only the learned who study natural history deeply, but also the common people among my readers, especially those who live in Cuba, in whom I want to root a love of science."[64] His inclusion of the phrase "especially those who live in Cuba" was a pointed rejoinder to Sagra's European collaborations. This interest in building affective relationships ("a love of science") between the common people and Cuban nature is even visible in the illustrations for *Memorias Sobre la Historia Natural.* While Sagra's Cuban animals were always shown separated from their environments, depicted placed against white backgrounds and floating in the center of the page, Poey's animals were sometimes placed within Cuban landscapes. Furthermore, these animals are often more emotive than Sagra's, further creating a sense of personal connection between the Cuban reader of his book and the Cuban nature depicted within,

The first volume opens with the chapter describing the almiquí rodent (plate 5). In the illustration for the chapter, the animal fills most of the page and its slightly scaly tail curls under his feet to fit into the frame. Frédéric Mialhe served as the lithographer for this illustration, and he accentuated the animal's soft fur with long streaks of alternating brown and pale yellow ink, producing a shimmering effect. The almiquí's body is surrounded by a soft halo of fur around its body. This emphasis on the visual and tactile qualities of the almiquí's fur draws the viewer's attention to the contrast between this luxurious pelt and its naked pink feet, tail, and underside. The almiquí is depicted in a lightly sketched environment, with a thick patch of greenery at the bottom left of the page and rocks scattered around its feet.

The animal's facial expression seems to be a mixture of shock and fear. Its mouth is open as if it is gasping and its small eyes are wide. Its limbs are bent slightly as if bracing for an attack from above, where Mialhe had placed the viewer. Its claws, which might have been depicted as dangerous weapons, appear instead as liabilities growing from the almiquí's toes at awkward angles. The viewer is confronted in this illustration with a terrified wild animal, as if we had stumbled upon the almiquí in the bush. This animal is seen in its natural environment, not floating against the page as Sagra's caiman did, and is understood partly through this environment—it seems almost to be hiding behind the clump of greenery to the left.

In the grisaille vignette of a pair of almiquís above the central animal's head, the two huddle together in front of more vaguely depicted greenery, one looking directly up at the viewer. Mialhe manipulated the strokes of the lithographic medium to give a sense of softness, even sentimentality. Each of the animals in this illustration seems to be a living specimen with realistic emotional reactions, and each is surrounded by tufts of grass. These visual elements combine to give the almiquí a vulnerable, sensitive quality that engenders sympathy. This emotional connection would have been one important technique to create a "love of science" in readers "who live in Cuba."

Another example of this emotive mode is seen on the page with various views of the *Hutia carabali,* also from volume 1 (figure 3.6). This page includes five vignettes of the rodent. The central view shows only its face, and the other four depict the animal climbing in a tree, hunting for food, or with a juvenile hutia for comparison. The detail of these vignettes—the animal's hand holding a fruit while it examines the food, the juvenile curled around its tail—seem intended to elicit an emotional response, even a tenderness, from the reader. The soft, brushy color lithography accentuates this quality. This page of hutias, like the illustration of the almiquí, was created by Poey's close friend Frédéric Mialhe.

Poey's decision to enlist his friend in creating the illustrations for his book might be seen as somewhat surprising and worth examining in more detail. Poey was entirely capable of creating the illustrations himself, and Mialhe was best known at the time not for scientific illustration but for his series of lithographed Cuban landscapes and cityscapes, published as *La Isla de Cuba Pintoresca* and *Viaje Pintoresco al Rededor de la Isla de Cuba.* These picturesque, celebratory views of Cuba were beloved by elite white Cubans and continue to be used today as rosy-hued celebrations of the island in the mid-nineteenth century. At the time that Poey asked him to work on this natural history volume, Mialhe was well known for images that sparked a warm, patriotic love for Cuba in the hearts of many elites on the island. Poey's choice to commission his friend can be explained by the argument that this was exactly the feeling he was also trying to elicit in his natural history publications.

Throughout his *Memorias Sobre la Historia Natural,* Poey celebrated the knowledge of everyday Cubans over that of professionalized Europeans and sought

Figure 3.6. Frédéric Mialhe, "Hutia Carabali," 1851, lithograph. (From Felipe Poey, *Memorias Sobre la Historia Natural de la Isla de Cuba,* volume 1 [Havana: Impr. de Barcina, 1851]; Smithsonian Libraries)

to elicit emotions of tenderness, awe, and love for Cuban nature in his readers, through illustration as well as text. This affective, democratizing view of Cuban science, however, did not seem to include, in Poey's eyes, the majority of Cubans. The knowledge of black or brown people was seldom acknowledged in his book, the Jamaican scientist of mixed race Richard Hill being the only exception.[65] This suggests that Poey only considered the knowledge of white Cubans to be of scientific relevance, and likely that he only imagined this love being sparked in the breasts of white readers of his book.

Poey was not unique in refusing to cite natural historical knowledge held by brown or black Cubans. Many historians of science have called attention recently to the central role people of color played as informants and knowledge producers in the project of nineteenth-century science, especially as practiced in the Caribbean and the Americas, but such individuals were rarely explicitly acknowledged in texts from the period.[66] This exclusion was not a given, however, and several writers of the period made different decisions. Both Sagra and Poey implicitly racialized their ontology of Cuban nature as white only. They cited white individuals and never mentioned the labor, expertise, or knowledge contributed to their projects by non-white individuals.

This decision helped to solidify a white identity in Cuba, by establishing white people as the holders of knowledge about the nature of the island. Poey wrote his book not just during a period of rising racial tensions and debates on the part of white Cubans about immigration, but specifically after the intense moment of racialized revolt and retributive violence known as the Year of the Lash. He was arguing for exclusively (white) criollo access to natural history knowledge in the years after the professional organizations and political rights of Cubans of color were being severely curtailed. We should understand this decision as part of the wider push to reinforce white supremacy at a time of intense change. This is not to discount Poey's other fascinating choices and accomplishments with this complex and surprising work but to remind us of the real-world effects that seemingly unrelated choices in producing scientific publications had in creating the social structure of Cuba after the Year of the Lash.

The other relevant context to *Memorias Sobre la Historia Natural* was, of course, the omnipresent sugar industry that continued to expand exponentially in Cuba as the century continued. One aspect of this expansion that Cuban scientists, planters, and the public more generally were becoming increasingly aware of during the time that Poey was preparing his manuscript was the problem of deforestation. Sugar production caused deforestation not just to clear land to create more fields for sugarcane, but also because the cane juice had to be boiled for many hours to concentrate it enough for shipping, and this boiling required vast quantities of wood. The demand for wood to drive the machinery of the plantation continued to increase as new steam-powered equipment, all imported from Europe, became more and more prevalent during the 1830s and 1840s. Because of this insatiable demand for wood for sugar production, Reinaldo Monzote points out that the amount of land covered by forests fell by one-third in the Central District of Cuba and by almost half in the Western District just from 1846 to 1862.[67] A visiting scientist, Arthur Morelet, wrote in 1847, "the traveler who tours the island of Cuba . . . is amazed by the barrenness of the land, which rarely offers shelter from the heat of the sun. And yet this countryside was once covered with forests that agriculture is causing to disappear."[68]

Deforestation was so extreme by the 1840s that even groups that were quite friendly to sugar interests, including the Sociedad Patriótica, published books and articles about the problem. These included a reprinting of *The Destruction of the Precious Cuban Forests and the Necessity of Replacing Them,* by the Conde de Mopox, in 1843.[69] It also included José Ricardo O'Farrill's *Report on Forests,* first published in 1792 and reprinted in 1851, which raised alerts about the danger of excessive deforestation and the necessity to rationalize its use, and the article "Recovery of the Forests Which Are Consumed Annually as Fuel in the Ingenios," by José Pizarro y Gardín, in the *Memorias de la Sociedad* of 1846.[70]

In the face of the loss of Cuban forests, Poey returned repeatedly in *Memorias Sobre la Historia Natural* to Cuba's natural beauty and biodiversity. He discussed the island's "varied nature" that "feeds the ardor of naturalists."[71] Poey did not mention sugar at all in these pages, but the invocation of varied nature implied a dichotomy between it and the monoculture of sugar. Poey would have been extremely aware of the deforestation being discussed in the publications of the Sociedad Patriótica, and in his book he chose to celebrate the biodiversity that would be lost as this deforestation continued.

We can even understand the visual structure of the illustrations of *Memorias Sobre la Historia Natural* as an implicit reference to this biodiversity. Across both volumes, the illustrations are defined by density, variety, and complexity, both in individual illustrations and across the illustrations more generally. The illustrations in Poey's book move from, for example, fish to shell to insect to mammal to fish to insect again. Poey also made use of a wide array of visual techniques ranging from the painterly, naturalistically rendered almiquí that opens the illustrations to flattened mathematical diagrams of butterfly wings. Nearly every time the reader turns a page there appears a new way of organizing visual information, as well as often a new type of flora or fauna. This diversity of visual styles gives the modern reader a sense of a multifaceted Cuban nature that is ever changing and unknowable in its totality.

Furthermore, where Sagra's illustrations show individual specimens floating against vast white spaces of the page, Poey filled every inch of each of his pages with dense detail and different views of the species being depicted. A typi-

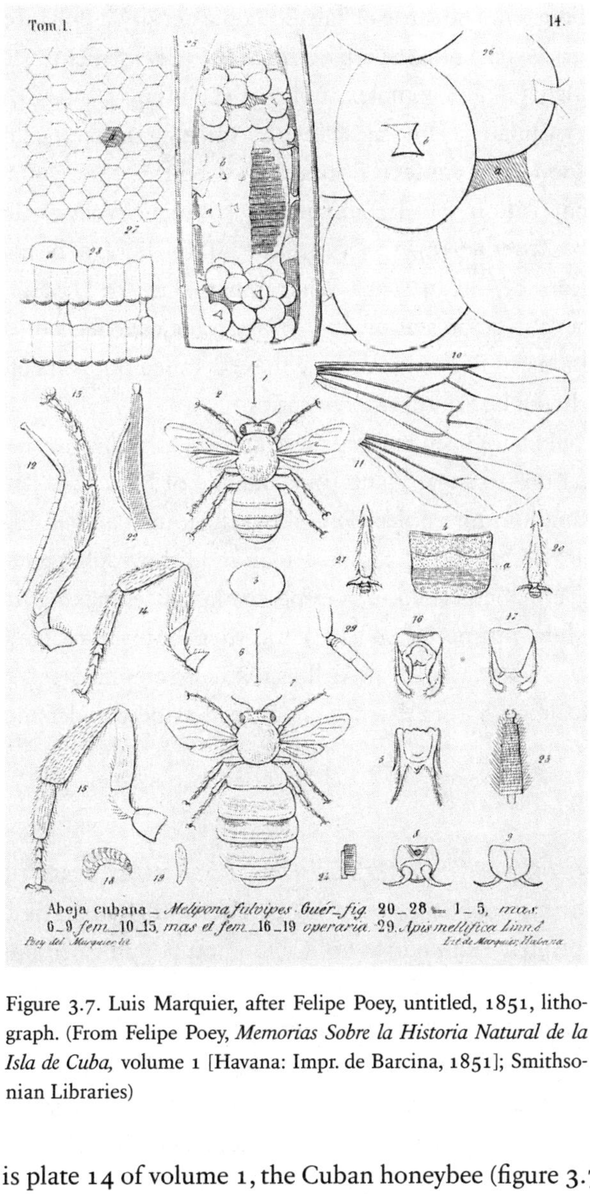

Figure 3.7. Luis Marquier, after Felipe Poey, untitled, 1851, lithograph. (From Felipe Poey, *Memorias Sobre la Historia Natural de la Isla de Cuba,* volume 1 [Havana: Impr. de Barcina, 1851]; Smithsonian Libraries)

cal example is plate 14 of volume 1, the Cuban honeybee (figure 3.7). This page is covered with linear black-and-white illustrations ranging vastly in scale and subject. It includes a representation of the cells of the beehive at actual size at the upper left, an entire hive inside a cross-section of a branch of a júcaro tree in the top center, highly magnified details of an extremely tiny part of the insect's mouthparts known as the ligula, and a circle indicating a cross-section of the insect's abdomen. The sense of kaleidoscopic density is enhanced by the high visual emphasis given along the top of the page to hive structures, each of which feels increasingly full and complex, and overwhelming.[72]

For Poey, Cuban nature was defined by variety and complexity, and needed to be understood in its context. Poey emphasized this variety not just through his dense compositions but by juxtaposing different types of illustrations in adjacent pages. Together, this variety was meant, I believe, to be contrasted with the monoculture of Cuban sugar. When many elites of Havana were raising concerns about the deforestation and other ecological effects of the expansion of the sugar industry, Poey was creating these detailed portraits of the biodiversity that was being lost in this deforestation.

Poey would have been making these anti-sugar arguments despite his own family connections to sugar plantations outside of Matanzas. There are other possible explanations for his decisions about the format and arrangement of the illustrations for *Memorias,* of course. Paper and lithographic plates were likely more expensive in Cuba than in Europe, for instance, making it expedient to place several different species on a page together, for instance. However, the fact remains that his book, visually as well as textually, emphasizes a dense and interconnected Cuban nature that we might read as best understood by criollos themselves.

"Faithful figures": Other Illustrated Scientific Texts of Cuba

The natural history publications of Felipe Poey and Ramón de la Sagra were not the only printed examinations of Cuban nature published during the mid-nineteenth century. At a time when printing technology on the island and the materials required for such publications were still rather limited, there was a surprising density and diversity of natural history texts being produced.[73] Why each of their authors invested heavily in these descriptions and depictions of Cuban nature and what each hoped to accomplish is the subject of rest of this chapter.

Falling somewhere between Poey's entangled criollo Cuban nature and Sagra's atomized Europeanized Cuban nature are the illustrations created by the Spanish-born Juan Lembeye Lartaud for his *Aves de la Isla de Cuba,* published in Havana in 1850. Lembeye was born in Spain but lived in Cuba for several decades, studying local birds and corresponding with naturalists across the globe. The book includes twenty illustrations, all of which show John James

Figure 3.8. Juan Lembeye, "Lam. no. 6," 1850, lithograph. (From Juan Lembeye, *Aves de la Isla de Cuba* [Havana: Imprenta del Tiempo, 1850]; Smithsonian Libraries)

Audubon's influence strongly, with entire compositions often copied directly from Audubon's prints, though they are not produced at life size as Audubon's birds famously were. Lembeye depicted his birds in motion, often mid-flight, against simplified backgrounds, plants, and even other birds. This decision is visible, for instance, on the page showing several types of kites and warblers, plate 7 (figure 3.8). The different bird species represented perch together on lushly realized flowering vines, facing in different directions, seeming almost to look at each other, and overlapping with the vine and its leaves and fruit. The simplified backgrounds and emphasis on dramatic movement and social engagement between individuals are hallmarks of Audubon's illustrations. These birds hew closer to the *Picus viridis* illustration by Édouard Traviès, instead of Sagra's floating individual birds against white backgrounds.

Unlike either Sagra's or Poey's projects, Lembeye's *Aves de la Isla de Cuba* was not intended to bring together all of the aspects of Cuban nature in a single volume. Much of the scientific information about the birds in Lembeye's book was taken from d'Orbigny's texts, which were also a strong influence on Sagra. Lembeye cited both d'Orbigny and Sagra at length in his introduction to the publication.[74] However, Lembeye's decision to print his book in Havana and to focus on his first-hand encounters with living animals in the landscape of Cuba itself ("in the same localities in which the birds reside," as he stated) recalled Poey's project—Cuban nature for Cuban readers.[75] It shows the spectrum that existed between Sagra's and Poey's universalizing projects, points of which were taken up by scientists with more specific goals in mind.

Lembeye's project is also noteworthy because it demonstrates the appetite that Cuban popular culture had for illustrations that related to Cuban natural history. Several of the illustrations from this book found their way into Cuban popular culture, reproduced as cigar wrappers and souvenir ceramic plates.[76] These reproductions show the appeal of natural history imagery for middle-class urban audiences engaged in constructing a Cubanidad identity during the political shifts of the mid-nineteenth century.

Francisco Sauvalle's *Flora Cubana,* published in 1873, was also a view of one specific aspect of Cuban nature. It was part of a longer lineage of Cuban floras, most of which consist only of text lists of different species of plants found on the island, with a minimum level of description and none of the historical and social narratives that framed Sagra's text or the personal accounts of collecting that framed Poey's. The first of these floras were *Plantae Wrightianae e Cuba Orientali,* written by the German botanist August Grisebach in collaboration with Charles Wright in 1860. Grisebach wrote another flora in 1866: *Catalogus Plantarum Cubensium.*[77] Both of these florae were published in Leipzig, largely from specimens and reports that Charles Wright sent to Europe (hence his place on the title page). Sauvalle's *Flora Cubana* was an update to these previous versions, giving additional information about species found on the island in the same list format.

Sauvalle's first version of *Flora Cubana* incorporated some elements of the making of both Sagra's *Historia Física* and Poey's *Memorias Sobre la Historia Natural.*[78] It was written and published in Cuba by a North America–born scientist with decades of experience studying Cuban nature, as was Poey, but it also represented a collaboration between non-Cuban naturalists, similar to Sagra's books. Sauvalle was born in South Carolina to French parents but spent much of his life in Cuba. He worked closely with the American botanist Charles Wright on editing and collating the text for *Flora Cubana.* Wright, a student of the well-known U.S. botanist Asa Gray, had worked on many different expeditions to the Caribbean and Central America. He wrote very little under his own name but collected information and specimens that served as the basis for publications by other scientists.[79] Wright represented a generation of non-Cuban naturalists who, in the wake of Poey and Sagra, spent much of the 1850s and 1860s mounting expeditions to explore Cuban nature.

The fact that both Sauvalle and Wright were originally from the United States also hints at the growth of American interest in Cuba more broadly at this time. In the mid-nineteenth century, some of the plantation owners who deeply feared slave uprising in Cuba hoped that an alliance with or even annexation by the United States (where slavery was, of course, still legal) would help to protect them from these uprisings. Narciso López was one such planter, and he actually planned and led an invasion of Cuba by six hundred United States soldiers on May 19, 1850. López had significantly overestimated Cuban planters' support for his plan, though, and was quickly forced to abandon it. American presidents, however, remained interested in Cuba—in 1853, President Franklin Pierce offered to buy the island from Spain. Whereas Poey's *Memorias Sobre la Historia Natural* was related to a moment of increasing criollo calls for local political power, Sauvalle's *Flora Cubana* was shaped by a period of increasing American intervention. At the same time, however, Sauvalle's publication also formed part of a lineage of publications of Cuban flora produced by professionalized European and American botanists in the second half of the nineteenth century.

This lineage of text-based flora makes Sauvalle's decision to create a fully illustrated second edition of his Cuban flora in the mid-1870s all the more surprising. No records survive of why Sauvalle made this decision, not anything about the making of the one hundred watercolors now held by the New York Botanical Garden's LuEsther T. Mertz Library. Only the idiosyncratic images themselves reveal any clues as to Sauvalle's goals and inspirations for the project of creating an illustrated version of the flora of Cuba in the last years of the Cuban republic.

Even the artist of the illustrations held at the Mertz Library is not definitely known, though it's very likely that Sauvalle made them himself. Wright did mention Sauvalle "making pencil drawings . . . plain but faithful figures" over the years that the two men knew each other.[80] Though the watercolors were never translated into printed illustrations, possibly because Sauvalle died in 1879, they reflect the shifts in Cuban botany that had been driven by Sagra and Poey for the previous four decades.

The watercolors in Sauvalle's collection each show a single species of plant, just as Sagra's prints had. They vary widely in format. Some are quite sketchy, some highly detailed, some drawn in pencil only and others painted in full color. This is not entirely unusual for drawings that were meant to be turned into printed illustrations—many artists would include only the most cursory of details, with the idea that the printmaker would fill in the rest based on this shorthand. Still, the wide variety of techniques that Sauvalle (or the artist or artists he worked with) used in these *Flora Cubana* drawings might mean that he was experimenting with how best to show these plants to international audiences.

Almost all of the illustrations show a specimen at its actual size, laid across the page without particular regard for symmetry or orderly placement of leaves or petals—as in his illustration of a type of metastomacea (plate 6), with its lack of symmetry, the contrast between yellow and green leaves, and pressed-seeming flowers scattered down the page. The haphazard arrangement of the parts of the plant, especially the leaves, which are shown curving and overlapping instead of pressed against the picture plane, gives this painting a sense of immediacy. This shows a significant departure from Sagra's approach to illustrating flora, in which each specimen was centered carefully on the page and leaves and flowers placed on either side of the stem to create a roughly symmetrical composition. Sauvalle's choice to instead depict plants at actual size whenever possible, and without special emphasis on symmetry or a systematized format, makes his plants appear lifelike and individualized to the modern viewer. Sauvalle may have been thinking of other nineteenth-century botanical artists who worked at a 1:1 scale, such as Alexander Postels, whose *Illustrationes Algarum* was published in Saint Petersburg in 1840, or even Anna Atkins and her celebrated cyanotype impressions, also of algae.

The most interesting detail about Sauvalle's watercolors is that many seem to have been executed using a graphite rubbing from an actual specimen as a pattern. The rubbed graphite traces of individual leaves, including their veins, can be seen under several of the pen and watercolor outlines. In these watercolors, Sauvalle or his artist had created the rubbing and then traced the outline of the leaf in pencil, and then drew in stems, flowers, and possibly more leaves. This use of graphite rubbing to create an illustration is somewhat unusual—as anyone who has created a botanical painting might know, it would likely have been easier to draw a plant by hand than to build off a graphite rubbing in this manner.

There are several possible explanations for this use of the rubbing technique. Sauvalle (or his unnamed artist) may have been in a hurry and considered this technique more expedient than drawing each leaf himself. Sauvalle may have had doubts about his own drawing and painting ability—he was never formally trained as a botanical artist, and his linework and painting skill are shaky and somewhat awkward in most of the paintings. However, this incorporation of graphite rubbings in his botanical watercolors also hints at an interesting use of indexicality on the part of Sauvalle. When he used these rubbings to serve as the basis for his paintings, he may have been trying to link the illustrations in the most direct connection possible to the actual Cuban specimen. That is to say, a rubbing of a leaf might have been considered not just more accurate in its representation of the leaf, but to create a more unimpeachable chain of transmission of information from the specimen to the artist to the reader of the final book. At a time when many scientists were deeply concerned with losses during transportation of specimens, this more reliable chain of transmission would have had great appeal.

Jennifer Roberts discusses James Audubon's related use of tracing, direct transfer, and 1:1 scale in his *Birds of America.* Audubon often wrote about how he did not just draw the birds in his illustrations from actual specimens and at life size, but that he transferred exact measurements of different specimens onto the paper using specialized measuring tools to ensure that every detail of the painting was accurate. As Roberts argues, many of the objects that she examines registered the complications of their own transmission through this indexical transfer, what she calls "formal preprocessing."[81] This formal preprocessing of distance, explains Roberts, worked to guard against future loss. Sauvalle's drawings took this idea of "preprocessing" one step further, by transferring parts of the specimen directly onto the paper through rubbing and other transfer techniques.

This concern with transfer, this attempt to avoid a loss of information by using indexical prints, could point to Sauvalle's larger concerns with accuracy and with loss of information as objects circulated across the Atlantic. It may even represent Sauvalle's response to Sagra's publication, whose illustrations had been drawn and printed in Paris from specimens that were often decades old and damaged by their transatlantic journey and storage since. By linking his drawings to rubbings of actual specimens, rubbings that would have been done while the plants were freshly collected to preserve the vein detail shown, Sauvalle may have, like Poey had, been arguing that Cuban science should be performed in Cuba itself. He may have been making a visual argument that close ties to the island were necessary to truly understand its flora. This is largely conjecture, of course. We are forced to compare apples to oranges here—Sauvalle's preparatory paintings for printed illustrations to the final prints in Sagra's. However, it is quite possible that Sauvalle's unusual choice to base his botanical paintings on graphite rubbings was a response to the larger debates taking place between figures like Sagra and figures like Poey about which vantage point—Cuba or France—best served the creation of knowledge about Cuban science.

For his own part, after the publication of *Memorias Sobre la Historia Natural,* Poey spent most of the last decades of his life working on a massive ichthyology, a list of all fish found in the bays, rivers, beaches, lakes, and even caves of Cuba. He drew many hundreds of fish specimens and collated reams of information about these species from published sources and from his own observations and those of his correspondents. The project, which at one point totaled twenty volumes of text and thirty volumes of illustrations, was never published in full during his lifetime, and only selections from his drawings and text have been published since his death.[82]

These nearly unimaginable numbers on their own testify to Poey's ambition with this project, as well as explains his challenges in getting the project published during his lifetime. This scope also testifies to the continuing political import of the scientific text in nineteenth-century Cuba. Poey had personal and professional reasons for his project. He had spent his entire career telling the story of Cuban flora and fauna in locations ranging from his own teaching to his natural history museum to his *Memorias Sobre la Historia Natural de la Isla de Cuba.* Poey's *Ictiología Cubana* should be considered an extension of this long attention to Cuban biodiversity. However, the fact that Poey dedicated so much time, attention, and money to this ambitious project, even as the Cuban revolutionary wars began to rage around him, also hints at the symbolic and po-

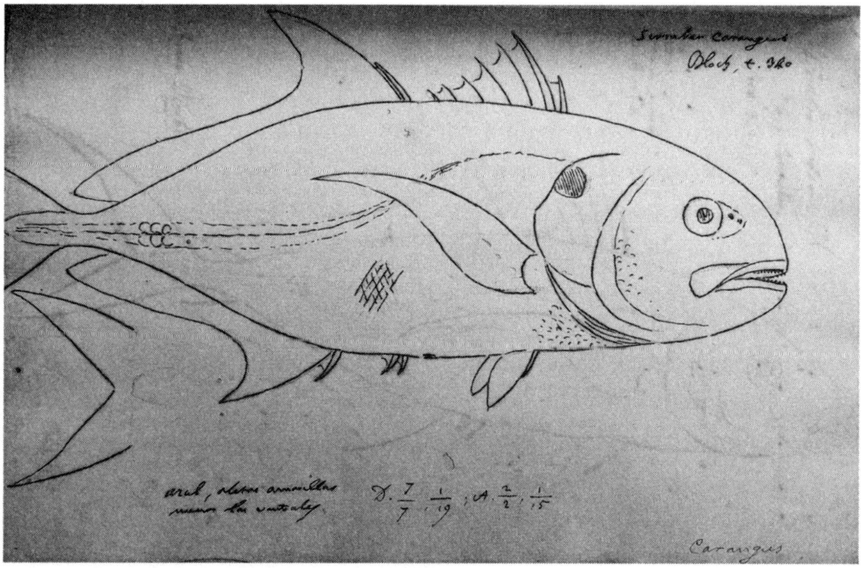

Figure 3.9. Felipe Poey, preparatory illustration for *Ictiología Cubana,* circa 1880, ink on paper. (Photograph by the author)

litical importance of fish in particular. Waterways connect locations, the ocean connects Cuba to other Caribbean islands and sites around the world, and Poey spent decades looking deeply into the inhabitants of the global networks of these waters. We might even think of the unfinished Ichthyology as the global turn at the end of a life dedicated to Cuban nature. It was also, however, a deeply Cuban project, an attempt to chart every part of one aspect of the island's nature.

The illustrations themselves also speak to Poey's particular attention to and hopes for this project. The hundreds of drawings in iron gall ink over graphite currently held by the Instituto de Literatura y Lingüística each show a different fish species laid flat on the paper, outlined with thick strokes of ink and with only a few rudimentary details included of scale size, pattern, or coloration. As seen in the example drawing of the fish in the Carangidae family (figure 3.9), the inked outline is somewhat broken and wavering at parts. In this example, the tail of the fish extends past the edge of the paper, and Poey drew the full tail directly below the fish's body. These thick, wavering lines are explained by Alderamo Romero's claim that Poey had polio as a child and experienced difficulty writing and drawing throughout his life. Romero also posits that because of this difficulty, Poey would lay the fish directly on the paper and trace around the outline, a theory that is upheld by several grease stains that remain on the pages held by the Instituto and the wavering graphite outline under the ink outlines.[83] If this theory of Poey's physical challenges is true, it would make the effort involved in drawing thousands of Cuban fishes for this Ichthyology even more impressive. This possible tracing of individual fish placed on sheets of paper also ties back to Sauvalle's graphite rubbings of leaves under the outlines of his flora illustrations, both likely attempts to imbue the resulting illustrations with a sense of authority and accuracy.

Selections from Poey's *Ictiología Cubana* were chosen by the Cuban exhibition committee to represent the island at the International Colonial and Export Exhibition in Amsterdam in 1883. This scientific text then became a symbol to international audiences of a Cuba that was struggling to free itself and define its own identity. The manuscript also gained some fame at this exhibition. King William III conferred on Poey the decoration of the order of the Dutch Lion, and the king of Spain distinguished him with the title Commander of the Order of Queen Elizabeth the Catholic.[84] Despite this success, the last years of Poey's life were spent trying to persuade different institutions to take the manuscript and drawings into their collection.[85]

The series of illustrated volumes described here that showcased Cuban flora and fauna, which at first might appear nearly devoid of political content, whether printed in Havana or Paris, consolidated scientific information about a range of Cuban species and were distributed around the world to scientists and amateur naturalists. They were intended, first and foremost, to convey scientific information about Cuban nature and to be placed in conversation with similar natural history volumes about other locations in scientific libraries. However, each of these volumes also reveals a range of political and social positions, and each reflected aspects of the larger context of their making. Furthermore, each volume presented knowledge of Cuban nature that was specifically situated in white experience.

The formal qualities of the illustrations of these volumes—Sagra's *Historia Física,* Sauvalle's *Flora Cubana,* and Poey's *Memorias Sobre la Historia Natural*—which at first might appear to convey only scientific information, substantiated the claims of ontology and knowledge production of the text. Sagra's illustrations depict species and specimens unrooted from their ecosystems, flattened against the blank page, just as his actual specimens were unrooted from their ecosystems and sent across the Atlantic. Poey's illustrations occasionally show animals in their environments and always use a densely packed page to imply the interconnected qualities of Cuban nature. Sauvalle, in his attempt to join a chain of scientific study of Cuba, forged the strongest visual links of this chain. He traced actual specimens onto his pages in order, perhaps, to create an indexical link between professionalized European botany and the Cuban flora itself. These projects all represented quite different arguments about what Cuban nature was and who could know it, and how. These arguments concerning race, ontology, and power were also mapped onto other representations of Cuban nature, especially landscapes.

4. "Make Known the Island of Cuba"

Images of the Landscape in and Around Havana

The production of natural history knowledge in Cuba in the mid-nineteenth century, as I've discussed throughout this book, was collaborative, contentious, and politically charged. This is true of all scientific knowledge production throughout history, though these qualities are almost never acknowledged. It is all the more surprising, then, that a lithograph made in Havana around 1842 does put some of these qualities into visual form. This is Frédéric Mialhe's "Fenómeno Natural en el Potrero de Dn. Manuel Lomba from Moreau" (plate 7). This print depicts two pairs of men, a massive rock formation, and an array of plants depicted in startling detail. One of the pairs of men consists of a white man wearing a hat and waistcoat pointing out the formation to another man whose back is turned to the viewer. The other pair consists of a black man wearing torn clothing and leaning toward another, seated, black man.

This print shows to the viewer—for perhaps the first time in the history of Cuban art—the scientific expeditions and the individual encounters with nature that had filled the beds of the botanical garden of Havana, the cases of Felipe Poey's natural history museum, and the pages of Ramón de la Sagra's books. Mialhe, in this print, depicted four men engaged in close looking at nature, and he also depicted the result of this close looking—the botanical knowledge that shapes the detail of the plants that surround the men. Mialhe had taken part in scientific expeditions himself. He traveled often with Felipe Poey, including into the Yumuri Valley and Matanzas, sketching while his friend collected plants and animals. One voyage became itself a case study of art and science's dependence on local knowledge—Poey and Mialhe got separated from the group and their local guide and were extremely lost for an entire day.[1]

Mialhe's print appears at first like many of the objects and spaces discussed previously, a celebration of specifically white knowledge and white ownership over the Cuban landscape. The two white men stand close to the center of the page, directly below the rock formation that draws the viewer's eye, placing them in a position of high visual importance in the scene. The postures of the figures in this print also establish a racial hierarchy—the white men stand while the black men stoop or sit. Furthermore, the clothing of the figures shows the white men to be wealthier than the black men, who are depicted dressed in rags and with bare legs.

However, this print also works against any easy claims to total white domination over Cuban nature. Mialhe's inclusion of the two black men, likely porters or guides, is itself unusual in prints from this period. Even more unusual is the placement of these black men in the exact center of the composition. Both of these formal decisions should be read as a statement about the botanist's and the artist's dependence on local black knowledge. As a reminder, all botanical expeditions and botanical collecting depended on labor—people needed to carry equipment and supplies, assist with preserving specimens and taking notes, and guide scientists safely through a wide range of landscapes to these specimens. In Cuba, many of these guides and laborers would have been black, and many would have been enslaved at local farms and plantations. Just as the botanical garden was made possible only through the labor and knowledge of enslaved black people, so too were the collections held in museum cases and printed onto pages. Mialhe represented the landscape of Cuba in this print as a site where different racial groups claimed knowledge and negotiated power. His close attention to the ecology of this landscape—to the plants that surround the figures and the stones—also hinted at the political importance of this ecology at a moment when the ecological catastrophe of the sugar plantation system was becoming ever more apparent.

In addition to spaces of natural history and encyclopedic scientific productions, the mid-nineteenth century in Cuba was also defined by the production of hundreds of prints depicting the Cuban landscape like "Fenómeno Natural en el Potrero de Dn. Manuel Lomba from Moreau." These prints were marketed both to Cubans and to audiences around the world. Much like the printed scientific treatises celebrating Cuban biodiversity discussed in the previous chapter, these landscape prints showed a lush, flourishing Cuba where a naturally abundant nature had been tamed and transformed into profitable agriculture and neatly delineated picturesque vistas.

In this chapter I examine three major lithograph projects of Cuba's mid-nineteenth century—Frédéric Mialhe's *La Isla de Cuba Pintoresca,* produced 1839–1842, the same artist's *Viaje Pintoresco al Rededor de la Isla de Cuba,* 1847–1848, and Eduardo Laplante's prints from *Los Ingenios,* from 1855–1857. Both Mialhe and Laplante were French printmakers who came to Cuba and aligned themselves with the members of the Sociedad, the plantation owners, and other members of wealthy white Cuban society. Both produced sweeping sets of prints that manipulated picturesque conventions of the period to celebrate the Cuban landscape and to present a narrative of progress over this landscape.

Although important scholarship has been done on each of these artists and their major Cuban projects separately, few authors discuss their output in conversation.[2] Further, almost no scholarship has been done on the relationship between these sets of lithographs and their wider social contexts, including the rise in elite interest in science and contemporary debates around the future of agriculture, sugar, and slavery in Cuba that I have traced. All three of these landscape and map print projects also worked to reinscribe white elite power over what was understood as the growing threat of Cuba's black population, similarly to what the botanical volumes and spaces discussed previously had done. Because of their form, however, they were able to make these visual arguments to elite white audiences beyond the scientific audiences of the volumes discussed in the previous chapter.

Landscape panorama prints, atlases, and maps in particular should be understood to have many levels of rhetorical qualities and political claims. As J. B. Harley reminds us, "the surveyor, whether consciously or otherwise, replicates not just the 'environment' in some abstract sense but equally the territorial imperatives of a particular political system."[3] Many scholars have also examined the political claims embedded in views of the landscape, especially sets of lithographs based on the picturesque format that were collated into volumes and shipped around the world, as Mialhe's prints were. Jeffrey Auerbach refers to this highly systematized way of representing the landscape as "a single formulaic lens" that was used in British landscape printmaking, for instance, to flatten local differences in far-flung imperial locations and to bring these locations more effectively into a shared imaginative space.[4] All of the artists that I discuss in this chapter consciously chose a highly systematized mode that was often used to naturalize and reinforce political power, especially imperial power, from among many different modes of landscape representation circulating across the globe at this time. Each of these prints should be understood as not just a window into a landscape but as a constructed political argument about this landscape. In this way, these landscape prints are quite similar to the botanical illustrations discussed in the previous chapter—objects that, thanks partly to their apparent neutrality, were able to encode complex arguments about knowledge, ownership, and futures of the island.

The prints of Cuban landscapes either denied or naturalized the ecological destruction of their time. The makers of these prints, who were members of the elite scientific circles discussed throughout this book, were highly aware of the increasing scholarship on how environmentally destructive the plantation system was, and specifically the deforestation that this system had caused across the island. These prints, then, should be understood as reactions to this destruction, reactions that ranged from the ecocritical to the celebratory.

"Rich in beautiful views": Frédéric Mialhe in Cuba

Lithography was first invented in Europe in 1796 and introduced in Cuba by 1822, but it was only in the 1830s that it truly became popular on the island. Involving drawing on a stone with hydrophobic ink instead of carving into the stone, lithography rapidly became quicker and cheaper than previous printmaking methods like engraving and etching. Zoila Lapique Becali has done excellent scholarship tracing the rise of the various presses in Havana in particular, especially on how these presses were connected to the Sociedad Patriótica.[5] One of these presses was run by the French-born Francisco Miguel Cosnier and the Martinique-born Alejandro Moreau de Jonnes, who came to Cuba in 1837. The two applied to the city's Captain General for permission to import new, advanced printing equipment in 1837, and the appeal was passed to the Sociedad Patriótica. The Sociedad granted permission, asking in return that the press be named after them. A subscription was taken to pay for the press (Felipe Poey was among the subscribers).[6]

Lithography, then, was a new technology with associations with cutting-edge European innovation that was quickly adopted by the Sociedad, which at the time was the most powerful conglomeration of sugar money and white elite privilege in Havana. The result, the Imprenta Litográfica de la Real Sociedad Patriótica, was inaugurated and celebrated with an announcement in the newspaper *Diario de la Habana* in August that brought together art and science, patriotism, and progress. While Havana was "progressing so rapidly towards its own expansion," the announcement stated, it still lacked "one of the indispensable elements in any country advanced in civilization": a lithographic press.[7] What should interest us, given the different modes of exploring and depicting Cuban nature that the Sociedad and associated individuals were engaging in during this time, is that this immense amount of wealth, privilege, and cutting-edge technology immediately went toward producing prints of the Cuban landscape.

It was probably Moreau who invited Frédéric Mialhe, who had been born in Bordeaux in 1810 and had had some success with his lithographic views of Mexico and the Pyrenees, to come to Cuba.[8] Mialhe arrived in late December 1838, and just four months later the press announced its intention to produce an extensive set of views of Cuba based on his drawings. Mialhe's publication *La Isla de Cuba Pintoresca* was the first major project undertaken by the press under the auspices of the Sociedad, and the Sociedad seemed especially invested in the success of the project. The advertisement for Mialhe's prints hinted at this convergence, celebrating "this truly national enterprise which will make known the island of Cuba and the beauties that it contains." The announcement went on to say that "the island of Cuba, [though] rich in beautiful views, remains unknown by lovers of nature: the intellectuals of Europe also turn their eyes towards this island of the Antilles and ask for its marvels to be revealed. . . . And while continental America has been visited . . . by travelers from all nations . . . it is time for Cuba to occupy its place in the admiration of the world."[9]

From the beginning, then, *La Isla de Cuba Pintoresca* was directed toward international audiences ("the intellectuals of Europe") as well as Cuban ones and was explicitly meant to be a celebratory project. Mialhe's lithographs were released in sets of four, with forty-eight released in total. Felipe Poey served on the committee from the Sociedad Patriótica that visited the press to report on the progress of the series. He wrote approvingly that Mialhe had "given us the best models of Cuban landscapes."[10]

Mialhe's approach to this landscape was discussed above, in his surprising and complex composition of "Fenómeno Natural." His representation of the Cuban landscape as a site of interracial sociality and ecological interdependence is also visible in the print "Baños de San Diego," which was likely published near the end of the project, in 1842 (figure 4.1). This print shows the titular calm pool of water, surrounded by massive stones and tangles of green-

Figure 4.1. Frédéric Mialhe, "Baños de San Diego," 1839–1840, lithograph. (From *La Isla de Cuba Pintoresca* [Havana: Litografía de la Real Sociedad Patriótica, 1839]; New-York Historical Society)

ery depicted in careful detail. Three black figures gather at the bottom of the composition, talking and looking out toward the pool, while a multiracial group arrives at the top. This group includes white people in tattered clothes as well as wealthy white men with top hats and an injured black person being carried on a litter by two black men. The print seems to celebrate the coming together of these groups at the spring. By centering the three black figures at the bottom, however, Mialhe may be referring to the fact that these natural springs are popularly thought to have been discovered in the seventeenth century by an enslaved black man, Taita Domingo.[11]

In preparing the final version of this lithograph, Mialhe made some changes to an earlier version that had originally been drawn by his printer Moreau for *El Plantel*, a Havana-based general interest periodical (figure 4.2). In Mialhe's version, he increased the density of the greenery on the left side of the print, added in more recognizably tropical plants such as palms, and gave additional visual emphasis to the man carried on the litter in the center. These changes served to situate the print more firmly in the specifically Cuban countryside, rather than a generically foliaged site, and to draw attention to the themes of vulnerability and interdependence that the litter invokes.

Figure 4.2. Alejandro Moreau, "Vista de los Baños de San Diego (Cuba)," circa 1839, lithograph. (From *El Plantel,* circa 1839; Beinecke Rare Book and Manuscript Library, Yale University)

Furthermore, this is a print about human/non-human relationality in the landscape. The figures are enmeshed in the plants around them, and these plants were given additional visual prominence in Mialhe's final version of the print. This enmeshing, in turn, might have been understood by the viewer of the print to enable the benefits that these figures gained from this landscape. Baños de San Diego was a celebrated and well-known site for healing waters. The mesh of plants surrounding the figures also may have been a reference to a Cuban ecology that retains its power through its biodiversity. It is possible that Mialhe intended this print to be a contrast to the deforestation that was taking place across the island at the time as a result of the expansion of Cuban sugar. Throughout *La Isla de Cuba Pintoresca,* Mialhe called attention to botanical detail and unique natural phenomena in conversation with the productive landscape, to black knowledge alongside black labor.

Many of Mialhe's prints include black figures to gesture toward black Cuban knowledge, as in "Baños de San Diego" and "Fenómena Natural," but other prints from the series reinscribe and celebrate the plantation slavery that helped pay for the entire series. "Cafetal la Ermita en las Lomas del Cusco," for instance, of a coffee plantation in the Pinar de Rio province in the north of Cuba, presents a neatly gridded plantation surrounded with wild greenery in the foreground of the print and hills covered with greenery behind (figure 4.3).

Figure 4.3. Frédéric Mialhe, "Cafetal la Ermita en las Lomas del Cusco," 1839–1840, lithograph. (From *La Isla de Cuba Pintoresca* [Havana: Litografía de la Real Sociedad Patriótica, 1839]; New-York Historical Society)

The conical shapes of the hills mirror the shape of the plantation building to the right of the print and the mounds of coffee in the centers of several of the fields. The bodies of the black enslaved laborers of the plantation were reduced to similarly repeated shapes in the fields on the left of the print. This print thus presents an image of de-individuated figures and alienated labor. It also naturalizes the plantation by emphasizing how its parts visually echo the surrounding landscape. The echoes between the shapes of the plantation building, the mounds of coffee, and the hills make a visual argument that each of these elements was a natural part of the landscape of Cuba rather than the highly planned and industrially produced buildings and products that we know them to have been. This is a celebratory vision of the Cuban landscape that would find its full extent in celebratory prints of plantations in the coming years.

The success of Mialhe's *La Isla de Cuba Pintoresca* spawned several imitators, the best known of which is *Paseo Pintoresco por la Isla de Cuba,* published between 1841 and 1842. This set of lithographs was published by a competing firm, the Litografía del Gobierno, and included eighty-six small views of Cuba.[12] *Paseo Pintoresco* was never as popular as Mialhe's prints were, and the set remains largely forgotten today. This is in part because the artists who worked on the series, including Fernando de la Costa and L. Cuevas, were not able to replicate Mialhe's specificity of place—of Cubanness. As a critic at the time wrote, "we want to know: what are we looking at, what landscape is this, or what part of the island does it show?"[13] The rise of imitators of Mialhe's prints does speak, however, to Cuban and international appetites for views of the island's landscape.

In the years after Mialhe produced *Isla de Cuba Pintoresca,* he remained in Havana and continued to work on various print projects there. During these same years, the socio-political situation in Havana became significantly more strained. Racial tensions increased during the first half of the 1840s, culminating in military investigations of rumors of island-wide slave uprisings in 1845 and violent reprisals toward black and brown communities across Cuba. In the years between Mialhe's first set of prints and his second set, the political mood of Havana shifted significantly, and rampant racial paranoia and persecution were accompanied by an increased desire for surveillance, signaling, and racial segregation on the part of the wealthy white Habaneros who made up much of his audience.

These explosions of racial tensions help to explain Mialhe's turn toward genre scenes that emphasized race and class differences in his second major print project, *Viaje Pintoresco al Rededor de la Isla de Cuba.* Mialhe's previous publishers, Cosnier and Moreau's Imprenta Litográfica de la Real Sociedad Patriótica, had folded in 1845, so *Viaje Pintoresco* was published by Francisco Louis Marquier's Litografía Cubana, founded in 1846. The French-born Marquier's press became an important node connecting art and science in Havana in the 1840s. Litografía Cubana printed illustrations for atlases, drawing manuals, sugar production manuals, political treatises, and major scientific texts by Poey and by Juan Lembeye. *Viaje Pintoresco,* like Mialhe's previous project *La Isla de Cuba Pintoresca,* was purchased by subscription, and the set eventually included thirty prints.

A sample print from this project is "El Casero (Habana)" (figure 4.4). In the center of the print, a black peddler haggles with a black woman over a bird he is selling her, while two more black figures look on. Immediately behind this scene, a white woman stands behind the grating of a ground-floor window, her arms grasping the grating, while a man looks up at her. This is an urban scene showing individuals of different races interacting and acting out some of the stereotypically held views of these races—the haggling black peddler, the ensconced and pure white woman. These scenes of everyday life that emphasize

Figure 4.4. Frédéric Mialhe, "El Casero," 1847–1848, hand-colored lithograph. (From *Viaje Pintoresco al Rededor de la Isla de Cuba* [Havana: Litografía Cubana, 1847–1848]; General Collection Manuscript Miscellany, Beinecke Rare Book and Manuscript Library, Yale University)

these racial and class distinctions are known as *costumbrismo.* Although this genre might at first appear entirely separate from the botanical illustrations and scientific spaces discussed across the rest of this book, *costumbrismo* takes as its driving force a kind of scientific, classificatory gaze, making these prints relevant to this larger discussion. This was a different view of the mixed-race population of Cuba than was seen in Mialhe's print of the Baños de San Diego—in "El Casero," the groups that interacted and even helped each other in "Baños de San Diego" were instead depicted as separated and encased for observation.

The woman in the back, in particular, presses up against the grating of her window to make herself visible both to the man in front of her and to the viewer of the print, like a botanical specimen pressing up against the picture plane. Similarly, many of the black figures turn their bodies or faces toward the picture plane for full visual access, just like Sagra's plants and caiman pressed up against the picture plane. Mialhe lavished attention on their clothing and the other details of their material existence—the different cuts of the dresses worn by the white woman and the black woman, the objects in the baskets carried by the peddler. We are meant to examine each figure in turn, to compare and

Figure 4.5. Frédéric Mialhe, "El Quitrín," 1847–1848, hand-colored lithograph. (From *Viaje Pintoresco al Rededor de la Isla de Cuba* [Havana: Litografía Cubana, 1847–1848]; General Collection Manuscript Miscellany, Beinecke Rare Book and Manuscript Library, Yale University)

contrast them, and to come away with a sense of knowing them. This evokes the details of teeth, claws, and skull that are magnified along the bottom of Sagra's caiman print.

This racializing, classificatory gaze is also present in the most famous and frequently reproduced print from *Viaje Pintoresco,* "El Quitrín," in which a black driver rides on a horse that pulls the titular large-wheeled carriage, with three white women sitting inside (figure 4.5). Mialhe emphasized details such as the intricate decorations of the carriage, the top hat and jacket of the driver, and the hairstyles of the women. He also drew attention to the difference between the very dark skin of the driver and the pale skin of his passengers, a difference that was made even more visible when the prints were hand-colored. It is noteworthy that between and above these figures sits the Fuente de la Noble Habana, which had an idealized indigenous woman as its central figure. The social landscape produced through costume and visual characteristics in "El Quitrín" is significantly more rigid and hierarchical than that produced through the interactions between figures and their environment as in Mialhe's earlier prints. This strict hierarchy is amplified by Mialhe's choice to make the horse that the driver rides dark black like the driver's skin, troublingly equating the black man with the well-trained animal below him. The racializing views of Mialhe's post-1845 prints retain a cultural power today. Asiel Sepulveda has pointed out how images from Mialhe's prints are scattered across the urban landscape of Havana today, often invoking an imagined idealized version of the city and its racial types.[14] This is in part because of the density of convincing details that fill these prints and because of some sense of comfort that white viewers still might feel in images of racial classification and definition.

"Havana described": Panoramas and Maps Produced in the Mid-Nineteenth Century

Other printed objects that seem to have been particularly well designed to instill a sense of white Cuban identity via representation of the landscape were the maps and atlases produced in the mid-nineteenth century. This was a moment when members of the Cuban intelligentsia were deeply invested in depicting and describing their island—Poey himself began his career writing geography textbooks for schoolchildren. Although the variety of maps of Cuba produced during the nineteenth century, both by Cubans and by foreigners, is too vast to examine in detail, there are a few projects that overlap in productive ways with the primary objects of this study. In these projects, the political possibilities of Cuban nature that I discuss throughout this book were projected quite clearly onto the map form.

Beginning in 1848 Marquier also printed a series of panoramas of Cuban cities drawn by Leonardo Barañano and lithographed by Eduardo Laplante. These were made on a large scale (approximately fifteen by forty inches) and were meant to be hung on the wall.[15] One example of these is *Vista de Matanzas,* from around 1860 (figure 4.6). In this print, three small figures in the foreground look out together over a bird's-eye view of the city of Matanzas, which was one of the centers of sugar production in Cuba throughout the nineteenth century. The figures in the foreground are surrounded by plants depicted in visual shorthand, and palm trees and neatly tilled fields fill the middle ground. The city itself rises up in the background, with the center street of Matanzas cutting through the center of the view of the city, leading the viewer's eye back into the depicted space. Small black-and-white views of major architectural features of the city, including the Palacio de Gobierno and the military hospital, surround this central panorama.

Figure 4.6. Eduardo Laplante, *Vista de Matanzas,* circa 1860, lithograph. (Courtesy Cernuda Arte, Coral Gables, Florida)

Panoramic views of cities were quite popular across the globe in the eighteenth and nineteenth centuries, and this print follows the conventions of the genre closely, including the use of repoussoir figures (the small objects, trees, or human figures that are placed in the extreme foreground of the composition to increase the viewer's sense of the depth of field), vignettes (the small scenes in circular frames placed around the panoramic view), and the flattened fabric of the city itself rising up in the back of the depicted space almost as if it were on a large ramp. What is interesting in this case is the appearance of this genre itself—the fact that elite urban audiences in Cuba had expanded enough and had enough interest in celebratory views of space to support this project. There was a kind of flowering of panoramas in Cuba at the time—there are also examples of panoramas of Havana that are signed by artists J. Bachmann, Willmann, and Smith and Hill, and Frédéric Mialhe himself. This flowering of interest in the display panoramic print is partially explained by the cultural production of the previous thirty years that I have traced—the gardens, public spaces, and exhibitions that instilled a sense of proto-nationalist identity in elite white Cubans and introduced them to the idea of taking pride in a display of Cuban nature.

A second map, engraved in Paris in 1843 by L. Bouffard and printed by Lemercier, was bound into the geographic atlas at the end of the second volume of Sagra's *Historia* (figure 4.7). This map, which measures approximately twenty-

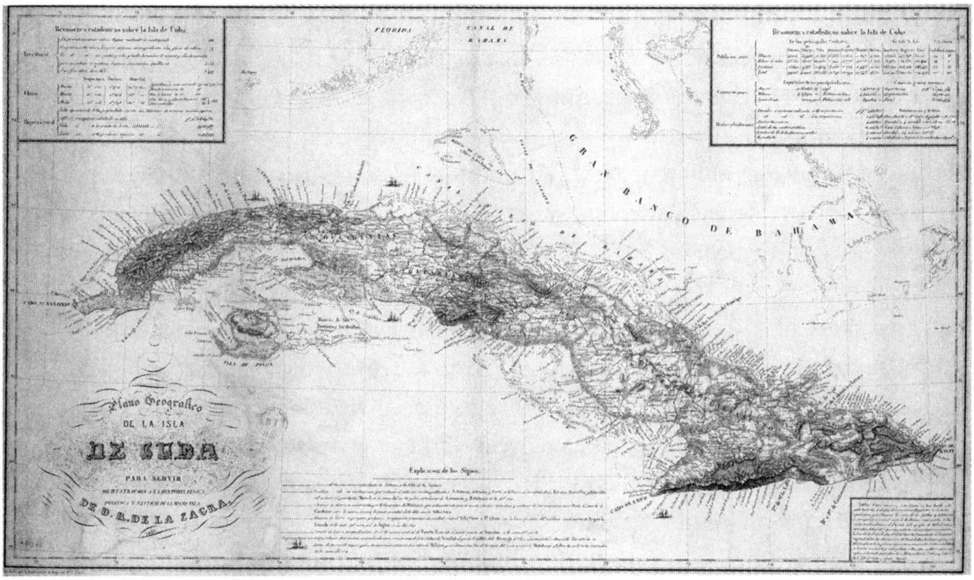

Figure 4.7. L. Bouffard, "Plano Geográfico de la Isla de Cuba," 1838–1861. (From Ramón de la Sagra, *Historia Física, Política y Natural de la Isla de Cuba, Introduction* [Paris: Maude et Renou and Bertrand, 1838–1861]; New York Public Library)

one by thirty-six inches, is the first major print the reader encounters in reading *Historia.*[16] In this map, every inch of Cuba is covered in a variety of marks. These include dense shading around mountain ranges, intricate intersecting lines representing roads, paths, and rivers, and circles, short hatch marks, and stippling dots representing plains and swamps. The map includes five different types of lines for the railroads that run back and forth across the island as well as small boats depicted at each of its major harbors.

These marks, and the fact that they cover the island so completely, implied that every kilometer of Cuba had been explored and mapped. This, of course, was not entirely the case in 1843. Through their density, the island appears three-dimensional on the page and almost looks like a kind of specimen itself, splayed against the picture plane. This map, then, seems to summarize Sagra's entire ambitious project. Just as was the case for Sagra's caiman and his clematis flower, all relevant data about this object was extracted, carried across the Atlantic, and then transformed by scientists and other professionals in Europe into visual information arranged on the page to allow for full visual access and understanding by the reader. A map tells the viewer how to understand a much larger space, including the island of Cuba itself. Sagra's map, like his entire

Historia Física project, showed viewers a Cuba that was defined and understood by circulation and white expertise. Just as that project sought to consolidate and claim all of Cuban nature, so too did his map seek to consolidate and claim the island itself.

An entirely different view of Cuba was presented in Rafael Rodríguez's *Atlas Cubano,* published in folios by the Imprenta Litográfica de la Real Sociedad Patriótica in 1841 and 1842.[17] The atlas includes seventeen large lithographed maps of cities and other sites of interest as well as a title page, dedication, and large sheet of historical, geographic, and statistical notes on the island. Both Sagra and Rodríguez, then, sought to create a definitive geography of rapidly changing Cuba, and both inscribed this geography in a larger historical framework. Both were Spanish-born employees of the crown—Rodríguez served as a lieutenant in the Artillery Guard.[18] However, the two men represented the geography of Cuba quite differently.

One of Rodríguez's maps, of San Carlos de Matanzas, is a good example of his project (figure 4.8). In this print, lithographed by the French-born artist and printer Francisco Cosnier, a topographic map showing the street grid of the city of Matanzas fills one side of the page, fading into farmland represented by cross-hatching, stippling, and small plant forms at the top and bottom of the page. Descriptions of the city, its history, and sets of statistics fill the right side of the page. The city seal of Matanzas appears at the top left, and the heraldry of the house of Ayllon at the top right. This is because the map was dedicated, in the text prominently placed just below the title of the map, and above Rafael Rodríguez's own name, to Don Cecilio Ayllon, the Spanish-born governor of the province of Matanzas from 1820 to 1832.

This print makes several arguments about how we should understand the geographic information it presents. It shows a single location in high detail, with the composition extending all the way to the edges of the paper itself. This implies the possibility of continuing the mapping off the edges of the paper. It further implies a sort of boundlessness of Cuban land, as opposed to Sagra's map of the whole island, entirely visible from afar and entirely self-contained. Of course, Sagra and Rodríguez were working on quite different scales and created quite different types of maps. However, it is noteworthy that Rodríguez did not include a print in his atlas showing the whole island of Cuba.

Notably, as well, Rodríguez's print explicitly inscribed geographic knowledge within a social network of rank and patronage. Each map in his *Atlas Cubano* was dedicated to a different political official, holder of a noble office, or

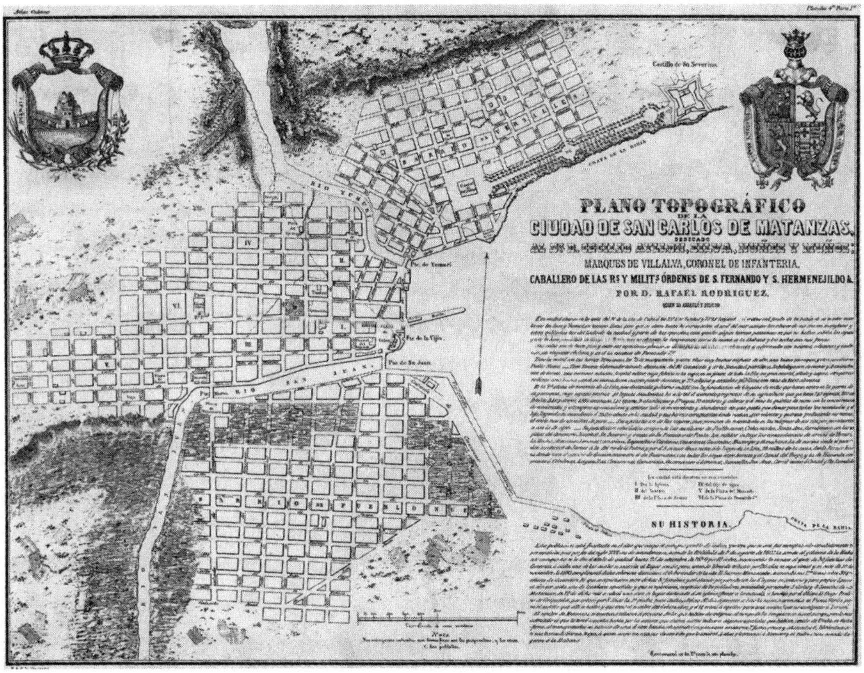

Figure 4.8. Rafael Rodríguez, "Plano Topográfico de la Ciudad de San Carlos de Matanzas," 1839–1842, lithograph. (From *Atlas Cubano: Blasones y Timbres que Decoran los Antiguos Títulos de los Soberanos de España* [Havana: Litografía de la Real Sociedad Económica, 1839–1842?]; Biblioteca Virtual de Defensa, Gobierno de España, Minsterio de Defensa)

owner of significant amounts of local land, often massive sugar plantations. In each dedication Rodríguez was careful to spell out the ways that his patron was connected to the realms of politics and military power—the dedication to Don Ayllon also named him as the marqués of Villalva and colonel of the infantry. These dedications are often visually reproduced by the inclusion of the patron's heraldry on the map itself, visually pinning down the geography of Cuba. Rodríguez, here, presented viewers with geographic knowledge structured and enabled by local social networks of powerful individuals.

Sagra's map imagines the island of Cuba as a single specimen, bounded and defined in relation to its neighboring land masses, while Rodríguez's *Atlas Cubano* emphasizes instead the dense and specific cartographic texture of local Cuban spaces. Where Sagra's map prioritizes cross-island movement and Caribbean and Atlantic circulation, and his *Historia* insists that the movement of objects to European metropoles was necessary to create knowledge, Rodríguez instead presented maps that were created by and inscribed with local elite Cuban social networks of patronage and land ownership. Furthermore, where Poey argued for knowledge of Cuba that was formed by the labor of non-elite Cubans, his contemporary and fellow geographer Rodríguez presented knowledge produced instead by networks of only the most elite intellectuals and politicians of Cuba. Again, this alternate proposal for how the land of Cuba should be understood was printed by the lithographic press that was partially owned and named after the Sociedad Patriótica. *Atlas Cubano* was, then, a project driven by and enabled by local elite Cuban networks which insisted on the primacy of these networks over the international scientific networks which Sagra celebrated.

"Most abundant source of the wealth and prosperity": Eduardo Laplante, Los Ingenios

Taken together, Mialhe's picturesque prints, the array of urban panoramas that were made in Cuba, and the maps that pictured the island presented urban life framed by nature, a diverse landscape that had been mapped and studied, and complex and negotiated social relationships. Many of these prints worked to deny the very real ecological disaster that also defined Cuban life during the mid-nineteenth century. This ecological disaster was a true example of what Sophie Sapp Moore, Monique Allewaert, Pablo F. Gómez, and Gregg Mitman have called the Plantationocene.[19] As discussed in previous chapters, both sugar plantation owners and the general elite educated public were increasingly aware of the ecological damage that sugar enacted on the Cuban landscape, and the deforestation of the island that accompanied its spread. Only one set of prints, however, rendered these changes visible. This is the thirty-five lithographs of *Los Ingenios.* The book was published in installments from 1855 to 1857 and included an introduction and text descriptions of twenty-five sugar plantations written by the criollo plantation owner Justo Germán Cantero. The illustrations were designed and lithographed by the French artist Eduardo Laplante y Borcou and printed by Havana-based, French-born Louis (Luis) Marquier.

A typical example of the plantation landscapes in *Los Ingenios* is "Ingenio Flor de Cuba" (plate 8). This print depicts the various buildings of the sugar plantation arranged across a smooth green plain with a wide road leading through them and a blue river stretching around the buildings to the right of the print. A few horse-drawn carriages, riding figures, and carts populate the road. Overseers on horseback are scattered throughout the buildings and fields, and the workers on the plantation themselves are barely visible as small dark shapes. The immediate foreground of the print features rolling hills and small repoussoir trees, and in the background the same green plains extend to a line of trees in the far distance.

To achieve this exact view of the plantation in physical space, the viewer would have to be suspended many feet above the plantation, looking down at it. This is similar to the bird's-eye view seen in the panorama of Matanzas discussed earlier, a view that many art historians have connected to rhetorics of power and ownership over the land pictured.[20] The road and the small horse-drawn carriages on this road lead the viewer's eye away from the hills and bushes in the bottom right of the print toward the neat and clean buildings of the sugar-making operation at its center. Our gaze is drawn by the visual elements of the composition away from the somewhat unruly nature of the foreground toward the harmonized and smooth plantation machine. This progression, I believe, would have been entirely intentional on Laplante's part, and the way this visual progression was structured recalls the history of Cuban sugar production taming and beautifying the land.

As Lapique Becali reminds us, the plantations discussed and pictured in *Los Ingenios* were largely chosen for their use of modern sugar technology.[21] These were truly industrial sites. Cantero reiterated this in the introduction to the project, stating that the ingenios pictured were "the most notable for the large scale of their products, those where some improvements or reforms of acknowledged utility were established, or those which shed light on some aspect of sugar production."[22] The illustrations include views of the entire plantation set within the landscape around it and interior views of the boiling house of the plantation. These were the large rooms where the sugarcane juice was boiled down into cakes of sugar. The visual qualities of these landscape prints show the changes to sugar production that they attempt to make sense of and how they differed from previous and contemporaneous descriptions of plantation landscapes. These visual qualities also relate to how Cantero described the project and the plantations themselves in the text of the book. Finally, these prints sought to actively deny and stave off the threats to Cuban sugar that were becoming more obvious while they were being made.

Eduardo Laplante had come to Cuba at the end of 1848 as a salesman of sugar refining machinery for the French company Derosne et Cail, and he continued to work as such throughout his time on the island. Cantero wrote that he met Laplante when he hired the artist to paint a view of his plantation.[23] The first announcement of their collaborative project, in the *Gaceta de la Habana* of May 3, 1855, emphasized the public importance of *Los Ingenios* as well as its aesthetic appeal: "We have the first delivery of the magnificent work that [is] . . . the most important that has come out of the presses of Havana . . . [which is of] real importance for the country and for the artistic interest that it offers."[24]

The phrase "real importance for the country" revealed the ultimate patriotic intent of the project. This claim of "importance to the country" was repeated twice in the announcement, indicating that the project was intended not just to celebrate the Cuban sugar industry, likely with the hope that the leaders of this industry would buy many copies of the sets of prints, but also to imagine the future of this industry, and the future of the island of Cuba itself. Laplante was responsible for both drawing and lithographing the illustrations of *Los Ingenios.* Lapique Becali adds that Ulises Marquier (the printer Louis Marquier's brother) and Santiago Martin y Martin, both born in France, assisted in the print run, including possibly in the production of the lithographs themselves, and that the type was printed in a nearby print shop, La Cubana.[25]

Nothing is known of Laplante's career or artistic training before he arrived in Cuba at the end of 1848. On January 4, 1849, he registered himself with the local authorities as "born in France, single, a businessman, 30 years old."[26] During his time on the island, Laplante also created several panoramic views of Cuban cities and countryside as well as illustrations for cigar packaging. As might be expected based on the format itself, Laplante's panoramic views, also printed by Louis Marquier, share the same high viewpoint over a wide swath of countryside that remains in focus far into the distance as the landscapes in *Los Ingenios* do.

For his own part, by the time he wrote *Los Ingenios,* Cantero owned eight plantations in total, including Ingenio Buena Vista near the city of Trinidad, west of Havana. He had studied medicine in Boston and Havana before marrying into a plantation-owning family. He took pride in hosting international visitors to Cuba, especially scientists, including Alexander von Humboldt and Juan Gundlach. Cantero was, therefore, highly attuned to a certain type of global network of scientific circulation.[27] He was likely hoping that his book celebrating the modern industry of Cuban sugar mills would find a similarly global audience and that it could also serve to elevate the island in the eyes of elite readers around the world.

One quality common to all the landscapes in *Los Ingenios* is Laplante's exaggerated perspective and the extreme shifts between the size of objects in the foreground and in the background. In "Ingenio Flor de Cuba," the road that occupies the center of the composition is quite wide in the foreground, taking up nearly a fifth of the bottom, but narrows rapidly as it moves toward the boiler room in the center left of the print. This forced one-point perspective is also seen in the proportions of the building at the far right, which stretches from left to right at an awkward angle. The exaggerated perspective is emphasized further by the bird's-eye viewpoint, extreme depth of field, and wide frame of "Ingenio Flor de Cuba." These strategies all ensure that the visual emphasis in this print is on the smoothed patchwork quality of the land and its enamel-like shades of greens and blues. The somewhat flattened plantation buildings have visual resonance with the patchwork quality and the color of the fields around them, making the entire plantation appear to be an inextricable part of the fabric of the land. Laplante's smooth fields of color of plantation building, plantation ground, and surrounding landscape, seen from above, appear to be one harmonious whole without disruption or conflict.[28]

Throughout the landscape prints of *Los Ingenios,* Laplante returned again and again to the basic visual grammar of bird's-eye view, plantation buildings interwoven with the flattened, expansive landscape, and exaggerated one-point perspective. Some of the later prints in the series, though, experiment with these building blocks in surprising ways. In "Ingenio Acana," for example, a train and train tracks fill the foreground of the print, parallel to its bottom edge (figure 4.9). The train, which we might take to represent power, prosperity, and speed, is moving toward the right of the page, and this movement is emphasized by the figures standing next to the tracks in front of the train and the smoke and steam coming from it. This steam and smoke echo the smoke emerging from the boiler chimneys of the plantation behind the train. Laplante thus created a visual connection between the modernity of the locomotive and of the plantation, both of which seem to be rushing into the future.

"Ingenio Buena-Vista" made Laplante's celebratory vision of Cuba's future most explicit (figure 4.10). In this print, the smooth planes of the sugar plantation and the neat boxes of its buildings are crowned by a golden sun with rays shooting out and illuminating the surrounding fields. The main house of the plantation owners stands on top of a small hill, with the rest of the plantation buildings spread out below. Just as in "Ingenio Flor de Cuba," the viewer's eye moves back toward this orderly scene of productivity along two different paths, one from right to left across the print and one from left to right. The former path leads straight back to the primary residence on the hill, while the second

Figure 4.9. Eduardo Laplante, "Ingenio Acana," 1855–1857, lithograph. (From Justo Cantero, *Los Ingenios* [Havana: Impreso en la litografía de Luis Marquier, 1855–1857]; Beinecke Rare Book and Manuscript Library, Yale University)

leads to the glowing sun. This makes a visual argument that the house, and by extension its inhabitants, the owners of the plantation, are themselves the sun rising over the landscape of Cuba, blessing this landscape and making it productive. This is particularly interesting as Ingenio Buena Vista belonged to Cantero himself.

As "Ingenio Buena-Vista" demonstrates, these were highly celebratory prints, which implied through their visual logic that plantation owners were all-powerful individuals who reshaped the land itself to their bidding. Antonio Benítez Rojo calls the visual logic of these prints "a kind of poetic panoply or myth."[29] A glowing review in the *Gaceta de la Habana* sheds some light on how they were received. "This publication is recommended to all people with good taste because of the particular care taken with its execution."[30] Good taste ("bueno gusto") was, as discussed earlier, a highly racialized concept at this time—a term that white elite Cubans claimed to further distinguish themselves from black and brown Cubans. This review hinted not just that the reception of *Los Ingenios* was positive, that people appreciated the high quality of its execution, but that white Cubans would have purchased the prints as a way to ensure their high social status and to consolidate their racial identity. All of the extant reviews of *Los Ingenios* are from Cuban newspapers and magazines—no informa-

Figure 4.10. Eduardo Laplante, "Ingenio Buena-Vista," 1855–1857, lithograph. (From Justo Cantero, *Los Ingenios* [Havana: Impreso en la litografía de Luis Marquier, 1855–1857]; Beinecke Rare Book and Manuscript Library, Yale University)

tion survives on how successful the project was with Cantero's hoped-for global audiences.

In the introduction to *Los Ingenios,* Cantero discussed meeting with Laplante and says that when he saw the "facility, taste, and exactitude of his drawing and his uncommonly high level of knowledge about our [Cuban] agriculture, we talked about how convenient a book showing the principal plantations of Cuba would be, and how it would make known the advances and efforts of the farmers to follow the universal march of progress and to generalize this knowledge in a country in which part of its inhabitants has so little inclination to travel . . . [perhaps due to] the love that inspires us for the town where we were born."[31] This "universal march of progress" was what Laplante had given visual form in the book's illustrations, especially "Ingenio Acana" and "Ingenio Buena-Vista."

Even though very little is known of Laplante's career or education before coming to Cuba, the fact that his plantation landscapes used a range of contemporary visual conventions, including bird's-eye views of the land, composition elements that drew the viewer's eye back into the depicted space, and repoussoir elements all show that he was familiar with representations of the landscape of his era. It is thus useful to spend some time discussing how his representations of Cuban plantations echoed and differed from contemporaneous depictions of plantations from elsewhere in the Caribbean region.

Many paintings and prints of the period represent the plantation as bucolic and Arcadian. The aquatints made in 1824 by James Hakewill depicting the great estates of Jamaican sugar planters are examples of this genre. Hakewill, born in Britain, had visited Jamaica in 1820–1821, when the prospects of the Jamaican sugar industry looked quite positive. However, the twenty-one aquatints he produced after his drawings, published as *A Picturesque Tour of the Island of Jamaica,* were printed beginning in 1824, after a major rebellion among enslaved Jamaicans and during intense legislative debates about reforms to the system.[32]

Like Laplante, then, Hakewill was producing images of Caribbean sugar plantations when the future of these plantations was increasingly uncertain. Like "Ingenio Flor de Cuba," Hakewill's "Montpelier Estates, St. James" also features a road that draws the viewer into the space of the print, toward the pictured estate (figure 4.11). However, Hakewill placed the viewer almost at ground level, whereas Laplante suspended the viewer dozens of feet in the air. The effect is that in the Hakewill print, the viewer is invited to identify with the two white figures on horseback in the foreground (and thus, close to the viewer's eye level). The grounds of the estate are not nearly so flattened in the Hakewill as in the Laplante prints, and these grounds rise up around the viewer as she moves through the space of the picture. As the eye moves back into the grounds of the plantation shown by Hakewill, viewers might imagine themselves as one of these white figures moving back through the space—a more intimate view inviting a more embodied sense of the space and implicating them in the power structures of the plantation.

As Tim Barringer writes, Hakewill's visual strategies borrowed heavily from J. M. W. Turner's picturesque compositions depicting the English and the Italian countryside, often directly referencing particular Turner prints. This served to reframe the "overseeing of estates with the leisurely enjoyment of a picturesque prospect" and turned the fraught plantation landscape into a calm and bucolic world.[33] Hakewill's print celebrated plantation life—like Laplante, he produced his prints mainly for the plantation-owning elite—but Hakewill's was a celebration of the physicality of the here and now, the grand buildings of the estate growing up from a fertile ground, where Laplante's were a celebration of the endless vista of plantation futures.

George Robertson's plantation paintings and the prints created based on

Figure 4.11. James Hakewill, "Montpelier Estates, St. James," 1825, lithograph. (From James Hakewill, *A Picturesque Tour of the Island of Jamaica* [London: Hurst and Robinson (etc.), 1825]; Yale Center for British Art)

them, which also depict Jamaica, emphasized a sense of fecundity and the plantation growing up out of the lush ground even more than Hakewill's prints do. *A View in the Island of Jamaica,* painted at the end of the eighteenth century, is one example (figure 4.12). In this image, a woman of mixed race stands before a kneeling black man who holds up a bunch of plantains from a large pile of produce in front of him for her inspection. Behind the two figures, cattle roam through a richly forested landscape marked by dappled light and a soft green glow. This scene was meant to be taking place on a plantation—in this case, that of Thomas Beckford, who had hired Robertson to produce depictions of his property—and yet there are no hints of labor, industrial production, or even systematized agriculture at all. Both of the figures in this painting, if they existed at all, would have been enslaved by Beckford, yet they are represented as self-sufficient and independently walking through the arcadian landscape. As Tim Barringer and Geoff Quilley have noted, "Robertson's image renders completely natural a system based on the mass relocation under duress of whole populations" and "represents the slaves as though restored to a supposedly innocent, original African state."[34] This idealized social dynamic is set in a verdant and rolling landscape, in which the figures appear to be merely strolling. The

Figure 4.12. James Mason, after George Robertson, *A View in the Island of Jamaica,* 1778, engraving. (Yale Center for British Art)

lush landscape spontaneously producing crops and the individualized human relationships it hosts are an entirely different way of seeing the sugar plantation from Laplante's industrialized fields and reified workers.

Laplante's illustrations of Cuban sugar plantations, in fact, bear more resemblance to representations of European factories, mills, and shipyards that were being produced at the beginning of the nineteenth century than they do to other depictions of the Caribbean plantation landscape. Many of these factory prints used a nearly bird's-eye point of view, vanishing points past the horizon, and sharp linear representations of the buildings of the site itself to elicit admiration from the viewer and to represent the factory as truly modern and productive (figure 4.13). These are all qualities that Laplante's illustrations share. Cantero's text repeatedly described the Cuban sugar plantation as a fully modern industrial site. The similarities that Laplante's illustrations shared with European representations of other industrial sites suggest that both author and artist were trying to position their publication within the wider context of modern global industry, to appeal to global audiences.

Prints such as "Ingenio Flor de Cuba" and "Ingenio Buena-Vista," then, are about the plantation system and the enslaved labor that made this system

Figure 4.13. George Endicott, after John Penniman, *Novelty Iron Works, Foot of 12th St. E.R. New York. Stillman, Allen & Co., Iron Founders, Steam Engine and General Machinery Manufacturers,* 1842–1855, lithograph printed in colors with hand-coloring. (Metropolitan Museum of Art, New York, The Edward W. C. Arnold Collection of New York Prints, Maps and Pictures, Bequest of Edward W. C. Arnold, 1954)

possible. However, while other plantation pictures refused to picture this labor, naturalized it through the use of picturesque conventions, or imagined workers engaged in peaceful leisure activities, the black laborers in Laplante's were often subsumed into the machinery of the plantation itself. Tiny black figures populate the yard of "Ingenio Flor de Cuba," for instance, but they are almost indistinguishable from the tools, barrels, and other objects that also fill this yard. In "Ingenio Buena-Vista," the enslaved plantation workers are depicted lined up in the yard for inspection, the repeated short vertical lines of their bodies echoing the lines of the trees and fences that surround them.

This reification of the black body in Laplante's illustrations is especially visible in the illustrations that show the interiors of plantation boiler houses. "Ingenio La Ponina" is a typical example of this (plate 9). This roof of the boiler house takes up most of the space in the composition, the rest of which is composed of highly geometricized elements such as the racks for the cakes of sugar in the foreground, the vertical and diagonal buttresses of the roof, and the long rows of boiling tanks at the left of the print.

These grids of sugar cakes in "Ingenio La Ponina" might remind us of William Clark's aquatint "Planting Cane," which he designed in Antigua in 1823

Figure 4.14. William Clark, "Planting Cane," 1823, hand-colored aquatint. (From William A. V. Clark, *Ten Views in the Island of Antigua: In Which Are Represented the Process of Sugar Making, and the Employment of the Negroes, in the Field, Boiling-house and Distillery* [London: Thomas Clay, 1823]; Yale Center for British Art)

(figure 4.14). Clark's print is also defined by strong grids, with the field that the sugar is to be planted in broken up into neatly defined sunken squares. This grid rolls across the bottom third of the print, and the figures of enslaved black laborers stand on top of and within it. Clark and Laplante treated the black bodies that they represented quite differently, however. The enslaved laborers in Clark's aquatint disrupt the grid of the planting field through the curves of their own bodies. All of the dozen or so workers depicted bend toward the ground, twist up with baskets on their heads, or turn to the side. Only the two overseers—one in the foreground carrying a whip and one in the middle ground wearing a top hat—stand perfectly upright. Furthermore, the workers themselves are also quite visually differentiated. They are depicted wearing clothing in a range of colors and patterns, wearing different headgear, and performing different tasks. This visual diversity contrasts sharply with the grid of the ground the enslaved individuals stand on.

What made Laplante's interiors unique was their total incorporation of the bodies of the black laborers into the smooth machinery of the boiling rooms. Boiling rooms were hellish places, incredibly hot and loud, running day and night. The threat of injury, whether by contact with the boilers themselves or when feeding the cane into the mill, was constant. Laplante's images emphatically deny this danger. Instead, he presented clean interiors that appear cool and comfortable through his use of deep shadows throughout. Dressed in white and depicted as small undifferentiated figures, the workers in "Ingenio La Ponina" push carts of sugar cakes in the foreground, oversee the operation in the middle ground, and operate the boilers in the background. They echo the upright buttresses and curved boilers, rather than interrupt the grid as in Clark's aquatint. Laplante seems to have worked to create the same sense of smoothly productive sugar machinery that the formerly enslaved writer Esteban Montejo described in his autobiography when he discussed the industrialized boiler room as nearly magically devoid of laboring bodies: "When I saw all those machines moving themselves at once I was astounded. They seemed to be going by themselves."[35] Both Laplante and Montejo seem to have been imagining the "plantation without slaves" discussed twenty years earlier in *Memorias,* though from entirely different points of view.

This vision of the boiler house as a seamless industrial machine related to technological changes in sugar production of the period. Most of the machinery represented in the Laplante interiors was produced by the French company Derosne and Cail (the same company that Laplante had come to Cuba to work for). The Derosne system had only begun to be widely adopted in Cuba in 1842 and used steam to heat the cane juice in vacuum-sealed containers instead of the open pans that Cuban ingenerios had previously used.[36] One of the selling points of the system, according to the manufacturers, was reduced labor demands and, through automatization, less possibility of human error. A brochure for the system celebrated how the design "facilitates operations, making them independent of the carelessness of the operatives; such that today the operative is *subjected to* the apparatus itself."[37] This subjugation of humans to machines is what was depicted in Laplante's interiors—a series of efficient machines dominating the space, with figures required only to monitor them. Of course, this automatization was largely aspirational. The new machines required significant expertise and training and still posed real dangers to workers.

The lack of any sense of threat, physical labor, or individualized emotion in Laplante's interiors is quite different, for example, from British depictions of iron forges from the previous century, though both genres celebrated industrial progress at a moment of transition. One such depiction is Richard Earlom's mezzotint, *An Iron Forge,* 1773, after Joseph Wright of Derby's painting of the

Figure 4.15. Richard Earlom, after Joseph Wright of Derby, *An Iron Forge,* 1773, mezzotint. (Yale Center for British Art)

same title from the previous year (figure 4.15). Where Wright, in his blacksmith paintings from the 1770s, emphasized the human drama of the scene through his signature intensely contrasting light and shadow, Laplante's interiors are lit with an even light that draws attention to the details of the machinery itself, and the expansive space of the interior. Where Wright maintained a close focus on just a few individuals, on their facial expressions and their muscles straining with the labor of the forge, Laplante's wide views minimized the presence of the black figures in his boiling rooms. Furthermore, none of Laplante's figures seem to be laboring in any real way. Given that Laplante's choice to depict the interiors of the sugar works was, itself, an unusual one, it seems possible that he knew of these previous British depictions of humans and machines. If that was the case, he seems to have carefully changed almost every aspect of these British scenes to reinforce a belief in absolute racial difference as well as to insist on the safety and ease of modern sugar machinery.

Along with the descriptions of each of the plantations pictured, *Los Inge-*

nios was also accompanied by its fourteen-page introduction by Justo Cantero, which was written and published after all the other sections had been produced, in February 1857. It reemphasized the goals of the entire project and the vision of Cuban sugar that had been laid out in Laplante's prints, which had been published over the previous two years. Cantero's text rarely discussed the laborers of the Ingenio, which corresponded to the visual deemphasis of these workers in Laplante's illustrations. He did write about the provisions necessary for laborers, including sustenance gardens, but only in passing.[38] Cantero's introduction, as should be expected, was certainly in favor of racialized slavery—at one point he dismissed activist claims about extreme punishment of enslaved people, saying that these were observed "only in very extreme cases"—and considered the enslavement of African people to be righteous, necessary, and entirely unremarkable in the maintenance of a successful ingenio.[39] Beyond this, however, the enslaved and indentured laborers of the plantation hardly registered for him, and he spent significantly more paragraphs expounding on, for instance, the utility of guano as a fertilizer than on the labor that made sugar production possible in the first place.

The second paragraph of the introduction praised sugar as "the most primordial and most abundant source of the wealth and prosperity of this beautiful and delightful country."[40] The vast majority of the rest of the text of the introduction continued to celebrate sugar and to predict its long and productive future in Cuba. Cantero wrote that the purpose of his project was "to promote by all possible means the improvements and reforms necessary for the good of the community."[41] Cantero's focus on sugar, of course, made sense—this was a book about sugar plantations, written by a plantation owner. His focus also, however, tracked a larger shift on the part of Cuban elites more broadly, toward equating this type of industrial sugar production with scientific progress more generally.

While much of the Sociedad's membership over the course of the first half of the nineteenth century had consisted of sugar plantation owners, the group had maintained a significant interest in scientific botany and small family farm agriculture, as demonstrated in projects ranging from the botanical garden of Havana to Felipe Poey's museum of natural history. By the late 1840s and early 1850s, however, the Sociedad had largely turned away from these projects to focus nearly exclusively on improving sugar technology. This was seen, for instance, in the fact that each month's issue of *Memorias* of 1849 contained a long article about improvements to sugar manufacture. Sugar technology continued to be discussed regularly in the following issues of *Memorias,* as the improvements pictured in Cantero's mechanistic boiling rooms spread slowly across the island.

Despite these technological improvements, however, and despite Cantero's celebration of the future of Cuban sugar and the laudatory tone of the images themselves, he was writing at a time when it was clear to plantation owners that racialized slavery could not be depended on indefinitely. Although sugar production in Cuba had been expanding rapidly since the end of the eighteenth century, its future was increasingly uncertain. Emancipation seemed likely to the plantation owners, especially after the Slavery Abolition Act ended slavery in the British Caribbean in 1833. Envisioning what a plantation without enslaved labor would look like had been a consistent theme in the *Memorias* since the mid-1830s.[42]

At the time that *Ingenios* was being published, some plantation owners believed that indenture could replace slavery. The Real Junta de Fomento y de Colonización had made arrangements in 1844 with British companies to begin importing indentured laborers from China to the island. This labor pool was made more official in Cuba with the passage of the "Regulations for the Introduction and Control of Spanish, Chinese, and Yucateco Colonists on the Island of Cuba" in 1854. In the year 1857 alone, some 8,547 indentured laborers arrived from China, having signed contracts to work for the sugar planters for eight years.[43] This system of indenture was not unique to Cuba, of course, and neighboring islands instituted their own versions of the practice, bringing workers from China, India, Madagascar, Mauritius, East Africa, and many other locations. Walton Look Lai estimates that in total, 536,310 immigrants came to the British West Indies alone between 1834 and 1918.[44] However, intense debates continued to rage about the system, and it was clear that it was not an indefinite solution.

The middle of the nineteenth century was also defined by the rise of beet sugar. Unlike sugarcane, sugar beets were able to be grown and processed in Europe itself, and grew better in cooler climates than in places like Cuba. Sagra wrote on the crop as early as 1827, and articles on beet sugar also appeared in *Memorias* in 1836 and 1849.[45] Cantero himself wrote that "brazen" beet sugar's rise had made it a rival of cane sugar.[46] By the end of 1857, the year Cantero finished *Los Ingenios,* the dropping price of beet sugar caused an economic crisis in Cuba and other sugar islands as plantations were forced to drive cane sugar prices ever lower to compete.[47] The crash led to increased borrowing and debt on the part of planters, speculation on the kinds of modernized mechanizations shown in Laplante's illustrations, and eventually the loss of many plantations to peninsular Spanish bankers when the terms of the mortgages could not be met.

Laplante's celebratory landscapes and text seem, in light of these shifts, to have been almost desperate attempts to convince the reader of the success of Cuban sugar despite mounting evidence to the contrary. They might be understood as dispatches from the final days of Cuban sugar hegemony and profits before a period of high debts and defaulted loans. An attempted economic reform to address the crisis began in 1866 but quickly stalled, and the Spanish government instead sent increased military forces to suppress the discontented planters. In response, a group of planters on the east of the island took up arms against the Spanish, leading to the Ten Years' War. This was the first of the Cuban wars of independence—the Republic of Cuba was established in 1902. Gillian McGillivray calls this entire process a "renegotiation of the colonial contract" by which planters had agreed to Spanish rule in return for the stability provided by military protection and a promise to extend legalized slavery in Cuba indefinitely.[48]

Despite these turns of fortune, industrialized sugar continued to maintain a major hold on Cuban economic life throughout the end of the nineteenth century and into the twentieth. Cuban sugar expanded and industrialized further with heavy investments from American corporate conglomerates in the 1890s and 1900s, leading the island to look more and more like the flattened plains of endless sugar production celebrated by Laplante's lithographs.[49] The Sociedad Económica, for its part, weathered the political storms of the 1860s and 1870s relatively unscathed, even as many similar associations emerged during periods of relatively relaxed conditions and then closed their doors during periods of stricter Spanish oversight.[50] A version of the Sociedad remains active today, though it has lost most of its political power and now mainly functions as a social group, organizing conferences, publications, and lectures.

La Dignidad Marquillas: *Images of Cuban Nature in Popular Culture*

Although the full scope of popular printmaking in colonial Cuba remains outside this book, one set of marquillas cigarreras produced by La Dignidad tobacco company from the 1860s speaks directly to the intersection of images of Cuban natural history and construction of white national identity and thus seems like an appropriate place to end this book. Marquillas cigarreras are small (five by three inches) chromolithographs (a form of lithography that used multiple stones to produce multi-colored images) that were produced by Cuban tobacco companies. They served as cigarette packaging for the domestic market, and so were distinct from (though of course related to) the better-known cigar box art that was printed for the international market. Several scholars, including Alison Fraunhar, have discussed the role that these popular prints played in producing and disseminating a Cuban national identity, a sense of Cubanidad structured by repeating visual motifs and references, in the years leading up to the wars of independence. Fraunhar goes on to demonstrate how this identity was racialized, with the figure of the mulata recurring on these cigarette prints as a site of desire and anxiety.[51] These fascinating examinations of the racializing imagery of these prints, however, have mentioned but not fully explored the role that images of nature played in this same construction of national identity.

A set of marquillas cigarreras held by the Biblioteca Nacional de España shows that images of Cuban nature, especially botanical illustrations, were also used on these popular prints and likely also became a site of production of a sense of Cuban national identity. These twenty chromolithographed cigarette wrappers each feature an illustration of a plant, flattened against a white background, as the central motif in their design. This central motif is surrounded, as is the case for other marquillas cigarreras, with a decorative border—in this case, featuring animals, more flowers, and an urn that appears to be made of white marble—and has the brand's symbol, a crowned woman with a cherub, on the right. These twenty images were meant to be collected and assembled into a card game, the rules for which are printed on the first of the prints.

In this set, then, the central part of the popular print of the marquilla cigarrera is taken up not by a celebratory landscape or a stereotypical figure, as in others of the genre, but by a reference to a botanical illustration (plate 10). Importantly, these are not functioning botanical illustrations—the labels on each are vague, titled something like "moss" rather than naming an individual species, and there is an insufficient level of detail to readily identify the flora or to perform a Linnaean comparison between two species. Even as they did not function as botanical illustrations, though, they referenced the format, with a single type of plant flattened against a white background and in the center of the frame. In this way, these marquilla illustrations have visual resonances with Sagra's clematis illustration discussed in Chapter 3, though again, with a much lower degree of scientific detail. This celebration of the form of the botanical illustration in the site that was used throughout this period to develop a sense of national identity among a popular Cuban audience indicates that this form would have had resonance for this audience. The form of the botanical illustration was likely used, alongside other popular motifs like the seductive mulata and the idealized sugar plantation, to develop this national identity. The association of this national identity with the form of the botanical illustration was made possible by the many projects that I have outlined throughout this book.

The mid-nineteenth century gave rise to a great number of ambitious and seemingly unrelated print projects that referenced Cuban nature or the Cuban countryside. Each of these projects, though, made a visual claim about the future of this countryside at a moment of heightened political tensions and rampant environmental destruction. Each project sought to envision this future and to disseminate this vision to diverse audiences across the island and around the world. Each project again reminds us of the political stakes of images of nature and representations of the land itself.

5. What Do the Political and Ecological Layers Mean?

In this book, I have examined the political and ecological layers of meaning of a wide range of spaces and objects. I argued that the botanical garden of Havana shifted from celebrating scientific botany to celebrating agricultural possibility to celebrating leisure, all the while helping to define white audiences through performances of leisure and contrast with enslaved labor. I examined public displays of Cuban nature, including the agricultural fair and the natural history museum, as spaces to define criollo identity and imagine a Cuban nature that was unique to the island itself and knowable through deep engagement with this island. I looked at illustrated natural history volumes as differentiated ways of extending this knowledge of Cuban nature globally, either made knowable through professionalized European expertise or by extra-imperial agents immersing themselves in Cuban nature. Finally, I discussed images of the Cuban landscape as either critiques of or celebrations of the Cuban plantationocene.

Throughout this study, I focused on the elite white people, mostly criollos, who made up the majority of the membership of the Sociedad and served as the drivers for most of these scientific projects. This focus on white elite creators and audiences enabled me to trace how natural history and Cuban nature were central to how these white elite Cubans were making sense of their political context and imagining the future of their island. This focus is, however, necessarily limiting, especially given how complex and diverse Cuban society was at the time of my study.

There are several ways of expanding this examination of how the visual culture of natural history drives political change. A study of free black and brown elite society in Havana would, I predict, reveal that these individuals also used images of nature to think through Cuban political futures. Several recent excellent histories examine how free Cubans of African descent in urban areas strategically positioned themselves within the rapidly shifting political power structures that I have examined.[1] Many of these focus on the Abakuá lodges and *cabildos de nación,* organizations that served as social networks, religious organizations, economic safety nets, and performing groups. However, much work remains to be done in exploring whether these groups were mobilizing representations of nature to make political claims. Given the rich symbolism that forms the core of many Afro-Cuban religions, it seems extremely likely that nature formed a key aspect of how the members of these groups represented themselves at this time of massive ecological change. A productive departure point could be the highly public choreographed processions that members of the *cabildos* took part in every year on Día de Reyes, the Feast of the Epiphany. These processions were often described as resembling giant snakes, and while participating in them, individuals wore costumes and carried banners and other objects, almost all of which referenced the natural world and the cosmology of the particular group.[2]

My examination could also be extended to include how women from Cuba and abroad located themselves within shifting discourses of nationalism and science. Several foreign women traveling in Cuba in the mid-nineteenth century published descriptions that shaped the international opinion of the island. These include the Swedish feminist reformer Fredrika Bremer and the United States–born abolitionists Mary Gardner Lowell and Julia Ward Howe.[3] Fredrika Bremer even created a set of twenty-eight watercolors during her visit.[4] Karen Robert discusses how Lowell's accounts of Cuba are gendered, with her attention to flora and her use of language referring to beauty and affect.[5] It would be productive to examine other accounts of Cuba written at the time for similarly gendered language and to map such language onto other contemporaneous descriptions of Cuban nature. In particular, the New Hampshire–born Nancy Kingsbury Wollstonecraft produced a fascinating set of illustrated volumes of natural history during a nearly ten-year residency in Matanzas, Cuba. Emilio Cueto has begun researching this manuscript and its 121 highly detailed illustrations, but much scholarship remains to be done on Wollstonecraft, her relationships with various scientific networks of the period, and the wider socio-political implications of her paintings.[6] Studying these women's texts and images would allow scholars to trace how they constructed their identity and how this identity shaped their depictions of the changing island of Cuba.

Much less has been written about Cuban women themselves, whether criolla, peninsular, free black, or enslaved. The primary surviving account from a criolla woman is that of María de las Mercedes Santa Cruz y Montalvo, the Countess of Merlin, who wrote a popular and insightful book about traveling through Cuba in the 1840s that has been discussed by Adriana Méndez Rodenas.[7] Additionally, several studies have been published examining the intersections of race and slave status in colonial Cuba.[8] Few of these, however, examine women's relationship with the changing Cuban environment amid the political changes discussed. One useful departure point could be that of Afro-Cuban women scientists, particularly enslaved midwives, who mobilized their scientific knowledge to claim political and economic power.[9]

In addition, my examination centers on Havana as a center of scientific, artistic, and literary production, but many other Cuban cities contained similarly complex networks, each with locally specific perspectives. Santiago de Cuba, for instance, was another important cultural hub in the nineteenth century, with a rich printing culture, its own Sociedad Económica, and a thriving social network of free people of color.[10]

Finally, the examination of how socio-political context shaped images of nature and how these images, in turn, affected socio-political possibilities can be extended to many additional places and periods. I chose Cuba for this study because both the political context and the environment itself were undergoing rapid transformation in the mid-nineteenth century. Many other islands in the Caribbean, however, were experiencing a different, but closely related, transformation during this period. So, too, were regions of North and South America and Australia as well as other sites around the world. And, of course, these periods of rapid transformation were not limited to the nineteenth century. A similar study could be undertaken of Haiti during its own sugar explosion of the eighteenth century, for example, of England during the Enclosure Acts, or of any number of sites around the world today.

It is only by bringing together a range of representations of and engagements with the natural world that the political layers of these objects and spaces can be fully known. The material culture of science and the visual culture of the environment does not just contain layered and complex meanings, but these meanings reveal each other when we take the most textured view possible of these cultures. This is more important than ever as we look at our own current ecological disasters, in which deforestation and sea level rise and global migrations, racialized violence, and genocide are inextricably interwoven. How we study nature, how we represent nature, and how we understand ourselves are all necessarily related, and these relationships are more urgent to understand day by day.

Notes

Abbreviations

AMN	Museo Naval, Madrid
AGI	Archivo General de Indias, Seville
AHNE	Archivo Histórico Nacional de España, Madrid
ANC	Archivo Nacional de Cuba, Havana
BNJM	Biblioteca Nacional José Martí, Havana
MNCN	Museo Nacional de Ciencias Naturales, Madrid
RJB	Archivos Fondos Real Jardín Botánico, Madrid

Introduction

1. *Memorias de la Real Sociedad Patriótica de la Habana,* vol. 1 (Havana: Oficina del Gobierno y Capitanía General, 1836), "Poblacion Blanca," 391–413.
2. *Memorias,* vol. 1 (1836), 428.
3. *Memorias,* vol. 2 (1836), 38.
4. *Memorias de la Real Sociedad Económica de la Habana,* vol. 1 (Havana: Oficina del Gobierno y de la Real Sociedad Patriótica, 1817), 40; emphasis added.
5. Louis A. Peréz, *On Becoming Cuban: Identity, Nationality, and Culture* (Chapel Hill: University of North Carolina Press, 1999); Gabriel B. Paquette, *Enlightenment, Governance, and Reform in Spain and Its Empire, 1759–1808* (New York: Palgrave Macmillan, 2008); Dalia Antonia Muller, *Cuban Émigrés and Independence in the Nineteenth-Century Gulf World,* Envisioning Cuba (Chapel Hill: University of North Carolina Press, 2017); Karen Kornweibel, *Writing for Inclusion: Literature, Race, and National Identity in Nineteenth-Century Cuba and the United States* (Lanham, Md.: Fairleigh Dickinson University Press, 2019); Javier Alvarado Planas, ed., *La Administración de Cuba en los Siglos XVIII y XIX,* primera edición (Madrid: Boletín Oficial del Estado and Centro de Estudios Políticos y Constitucionales, 2017).
6. Alexander von Humboldt, *Political Essay on the Island of Cuba: A Critical Edition,* ed. Vera M. Kutzinski and Ottmar Ette, trans. J. Bradford Anderson (Chicago: University of Chicago Press, 2011), 26.

7. In thinking through the role of material culture in asserting group identity in a multivalent harbor city during shifting political circumstances, I draw in particular from Edward S. Cooke Jr., *Inventing Boston: Design, Production, and Consumption* (New Haven: Yale University Press, 2019).
8. Ramón de la Sagra, *Histoire Physique, Politique et Naturelle de l'ile de Cuba,* trans. M. S. Berthelot, vol. 1: Introduction, Géographie, Climat et Population (Paris: A. Bertrand, 1838), fig. 1.
9. Ann Twinam, *Purchasing Whiteness: Pardos, Mulattos, and the Quest for Social Mobility in the Spanish Indies* (Stanford: Stanford University Press, 2015).
10. David A. Sartorius, *Ever Faithful: Race, Loyalty, and the Ends of Empire in Spanish Cuba* (Durham: Duke University Press, 2013).
11. Rebecca J. Scott, *Slave Emancipation in Cuba: The Transition to Free Labor, 1860–1899* (Princeton: Princeton University Press, 1985); Adriana Chira, *Patchwork Freedoms: Law, Slavery, and Race Beyond Cuba's Plantations* (Cambridge: Cambridge University Press, 2022).
12. Scott, *Slave Emancipation in Cuba;* Bonnie A. Lucero, *A Cuban City, Segregated: Race and Urbanization in the Nineteenth Century* (Tuscaloosa: University of Alabama Press, 2019); Chira, *Patchwork Freedoms.*
13. Isabel Mendizabal et al., "Genetic Origin, Admixture, and Asymmetry in Maternal and Paternal Human Lineages in Cuba," *BMC Evolutionary Biology* 8 (2008); José Barreiro, "Indigenous Cuba: Hidden in Plain Sight," *American Indian Magazine* (Winter 2017).
14. José Ramón Rodriguez Alonso, "Breve Reseña Histórica Sobre el Uso de Los Venenos de Escorpiones Como Medicamento Contra el Cáncer," *Revista Información Científica* 74, no. 2 (2012).
15. Sherry Johnson, *The Social Transformation of Eighteenth-Century Cuba* (Gainesville: University Press of Florida, 2004).
16. Consuelo Naranjo Orovio and Armando García González, *Raciscmo e Immigración en Cuba en el Siglo XIX* (Madrid: Ediciones Doce Calles; Consejo Superior de Investigaciones Científicas, 1996).
17. Louis A. Pérez, *Slaves, Sugar, and Colonial Society: Travel Accounts of Cuba 1801–1899* (Wilmington: Scholarly Resources, 1992), xv. See also Manuel Moreno Fraginals, *El Ingenio: El Complejo Económico Cubano del Azúcar* (Havana: Comisión Nacional Cubana de la UNESCO, 1964).
18. Fernando Ortiz, *Contrapunteo Cubano del Tabaco y el Azúcar* (Havana: Jesús Montero, 1940).
19. Rafael de Bivar Marquese, "Luso-Brazilian Enlightenment and the Circulation of Caribbean Slavery-Related Knowledge: The Establishment of the Brazilian Coffee Culture from a Comparative Perspective," *História, Ciências, Saúde-Manguinhos,* no. 4 (December 2009):

855–80; Rafael de Bivar Marquese, "Moradia Escrava na Era do Tráfico Ilegal: Senzalas Rurais no Brasil e em Cuba, c. 1830–1860," *Anais do Museu Paulista* 13, no. 2 (December 2005): 165–88; Marta Macedo, "Coffee on the Move: Technology, Labour, and Race in the Making of a Transatlantic Plantation System," *Mobilities* 16, no. 2: Mobile Labour (2021): 262–72.

20. Theresa A. Singleton, "Nineteenth-Century Built Landscape of Plantation Slavery in Comparative Perspective," in *The Archaeology of Slavery: A Comparative Approach to Captivity and Coercion,* ed. Lydia Wilson Marshall (Carbondale: Southern Illinois University Press, 2015); Dale Tomich, "The 'Second Slavery': Bonded Labor and the Transformation of the Nineteenth-Century World Economy," in *Critical Readings on Global Slavery,* ed. Damian Alan Pargas and Felicia Roşu (Leiden: Brill, 2018), 1326–49.

21. I also assume that the production and sharing of knowledge about nature took place in a much wider universe than that of the white, formally educated professionals upon whom I focus. Scientific knowledge also extended to knowing when to plant and harvest sugarcane, how to mix and rotate different crops in a provision garden to fulfill nutritional needs, and what ceremonies to perform to expel sickness from individuals. It included knowing which plants should be offered to which spiritual beings at which time of the day, knowing how long to soak beans before cooking, and knowing what wood is soft enough to carve but hard enough for a piece of furniture made from it to last for generations. Many of the networks of production and circulation of these various knowledges fall outside of the scope of this book. These networks were embedded within the projects that I write about, though, since these projects would not have been possible without this wider scientific community.

22. One of the most important scholars to propose racial anxiety as a driving factor of white Cuban society at this time is Consuelo Naranjo Orovio, especially in "El Temor a la 'Africanización'": Colonización Blanca y Nuevas Poblaciones en Cuba (el Caso de Cienfuegos)," in *Las Antillas en la Era de las Luces y la Revolución,* ed. José Antonio Piqueras (Madrid: Siglo XXI de España, 2005), 85–121.

23. S. J. Watts, *Epidemics and History: Disease, Power, and Imperialism* (New Haven: Yale University Press, 1997); Richard B. Sheridan, *Doctors and Slaves: A Medical and Demographic History of Slavery in the British West Indies, 1680–1834* (Cambridge: Cambridge University Press, 1985). See also John Rankin, *Healing the African Body: British Medicine in West Africa, 1800–1860,* 1st edition (Columbia: University of Missouri Press, 2015).

24. Useful overviews of the historiography of tropical medicine include Benedikt Stuchtey, "Introduction: Toward a Comparative History of Science and Tropical Medicine in Imperial Culture Since 1800," in *Science Across the European Empires, 1800–1950* (Oxford: Oxford University Press, 2005), 1–46; Juanita De Barros, Steven Paul Palmer, and David Wright, "Introduction," in *Health and Medicine in the Circum-Caribbean, 1800–1968,*

Routledge Studies in the Social History of Medicine 33 (New York: Routledge, 2009), 1–18; Warwick Anderson, "Disease, Race, and Empire," *Bulletin for the History of Medicine* 70 (1996): 62–67.

25. Ikuko Asaka, *Tropical Freedom: Climate, Settler Colonialism, and Black Exclusion in the Age of Emancipation* (Durham: Duke University Press, 2017), 14. See also Sonya Posmentier, *Cultivation and Catastrophe: The Lyric Ecology of Modern Black Literature,* The Callaloo African Diaspora Series (Baltimore: Johns Hopkins University Press, 2017); Katherine Adams, "Du Bois, Dirt Determinism, and the Reconstruction of Global Value," *American Literary History* 31, no. 4 (November 1, 2019): 715–40.

26. Michele Reid-Vazquez, *The Year of the Lash: Free People of Color in Cuba and the Nineteenth-Century Atlantic World* (Athens: University of Georgia Press, 2011); Aisha K. Finch, *Rethinking Slave Rebellion in Cuba: La Escalera and the Insurgencies of 1841–1844* (Chapel Hill: University of North Carolina Press, 2015); Sartorius, *Ever Faithful.*

27. This process has been especially well studied in the United States, in, for instance, Theodore Allen, *The Invention of the White Race,* Haymarket Series (London: Verso, 1994); Matthew Frye Jacobson, *Whiteness of a Different Color: European Immigrants and the Alchemy of Race* (Cambridge: Harvard University Press, 1998); Stephen Middleton et al., eds., *The Construction of Whiteness: An Interdisciplinary Analysis of Race Formation and the Meaning of a White Identity* (Jackson: University Press of Mississippi, 2016).

28. The history of tropical medicine and climatic anxieties of colonists is an exception to this oversight. See, for instance, Anderson, "Disease, Race, and Empire"; Dane Kennedy, "The Perils of the Midday Sun: Climatic Anxieties in the Colonial Tropics," in *Imperialism and the Natural World,* ed. John M. MacKenzie, Studies in Imperialism (Manchester: Manchester University Press, 1990), 118–40; Philip D. Curtin, *Death by Migration: Europe's Encounter with the Tropical World in the Nineteenth Century* (Cambridge: Cambridge University Press, 1989).

29. Naranjo Orovio and García González, *Racismo e Inmigración en Cuba en el Siglo XIX;* Armando García González, *El Estigma del Color: Saberes y Prejuicios Sobre las Razas en la Ciencia Hispanocubana del Siglo XIX* (Santa Cruz de Tenerife: Idea, 2008).

30. Sibylle Fischer, *Modernity Disavowed: Haiti and the Cultures of Slavery in the Age of Revolution* (Durham: Duke University Press, 2004).

31. Lorraine Daston and Peter Galison, *Objectivity* (New York: Zone Books, 2007); Horst Bredekamp, Vera Dünkel, and Birgit Schneider, eds., *The Technical Image: A History of Styles in Scientific Imagery* (Chicago: University of Chicago Press, 2015); Lorraine Daston and Elizabeth Lunbeck, eds., *Histories of Scientific Observation* (Chicago: University of Chicago Press, 2011); L. J. Jordanova, *The Look of the Past: Visual and Material Evidence in Historical Practice* (Cambridge: Cambridge University Press, 2012).

32. See especially Alina Payne, ed., *Vision and Its Instruments: Art, Science, and Technology in*

Early Modern Europe (University Park: Pennsylvania State University Press, 2015); Barbara Maria Stafford, *Ribbon of Darkness* (Chicago: University of Chicago Press, 2019).

33. For more, see especially Alfonso Quiroz, "Free Association and Civil Society in Cuba, 1787–1895," *Journal of Latin American Studies* 43, no. 1 (February 1, 2011): 33–64.
34. The name of the group, the name of its associated publication, and the way the volumes of this publication were numbered changed many times over the course of the nineteenth century. All citations to the publication put out by this group give both the year and the volume number to reduce confusion.
35. Christopher Ianini, *Fatal Revolutions: Natural History, West Indian Slavery, and the Routes of American Literature* (Chapel Hill: University of North Carolina Press, 2010).
36. Ann Laura Stoler, *Along the Archival Grain: Epistemic Anxieties and Colonial Common Sense* (Princeton: Princeton University Press, 2009).
37. Saidiya V. Hartman, "Venus in Two Acts," *Small Axe* 12, no. 2 (June 2008): 2.
38. Dell Upton, "White and Black Landscapes in Eighteenth-Century Virginia," *Places* 2, no. 2 (1984): 59.

1. "In Order to Dominate It"

1. "Informe de M. Espinosa Dirigido al Intendente del Real Jardín Botánico Sobre el Jardín Botánico de la Habana," September 11, 1796, I.9.4.1, Archivos Fondos Real Jardín Botánico.
2. On the botanical garden of Havana, see especially José González de la Peña Puerta, Antonio Ramos Carrillo, and Estaban Moreno Toral, "El Jardín Botánica y la Botánica Farmecéutica en la Habana del Siglo XIX," *Ars Pharmaceutica* 53, no. 3 (2012): 34–39; and Miguel Ángel Puig-Samper and Mercedes Valero, *Historia del Jardín Botánico de la Habana,* Theatrum naturae (Aranjuez: Ediciones Doce Calles; Consejo Superior de Investigaciones Científicas, 2000). On botanical gardens across the world, see Lucile Brockway, *Science and Colonial Expansion: The Role of the British Royal Botanic Gardens* (New Haven: Yale University Press, 2002); Donal McCracken, *Gardens of Empire: Botanical Institutions of the Victorian British Empire* (London: Leicester University Press, 1997); Londa Schiebinger, *Plants and Empire: Colonial Bioprospecting in the Atlantic World* (Cambridge: Harvard University Press, 2004), among others.
3. William Howard Adams, *Nature Perfected: Gardens Through History* (New York: Abbeville, 1991); Patrick Taylor, *The Oxford Companion to the Garden* (Oxford: Oxford University Press, 2008); Linda A. Chisholm and Michael D. Garber, *The History of Landscape Design in 100 Gardens* (Portland, Ore.: Timber Press, 2018).
4. Bleichmar, *Visible Empire.*
5. Miguel Ángel Puig-Samper, "Illustrators of the New World," *Culture & History Digital Journal* 1 (2012), doi: http://dx.doi.org/10.3989/chdj.2012.m102.

6. *Memorias,* vol. 1 (1817), 2.
7. Iris Wilson Engstrand, *Spanish Scientists in the New World: The Eighteenth-Century Expeditions* (Seattle: University of Washington Press, 1981), 14.
8. Puig-Samper and Valero, *Historia del Jardín Botánico de la Habana,* 73.
9. *Memorias,* vol. 5 (1818), 294–302, cited in Puig-Samper and Valero, *Historia del Jardín Botánico de la Habana,* 78.
10. *Memorias,* vol. 1 (1817), 420.
11. This is an interesting aside which should remind us that while scientific botany was frequently closely tied to networks controlled by imperial organizations, it was not restricted to these networks. Even the most "loyal" scientific botanists, including Espinosa, maintained extra-imperial networks of correspondence to acquire new specimens.
12. AGI, Ultramar, leg. 107, num. 35.
13. AGI, Ultramar, leg. 108, December 29, 1821, letter.
14. For more on the legal frameworks of this era, see especially Chira, *Patchwork Freedoms,* chapter 3.
15. Sartorius, *Ever Faithful,* 1.
16. AGI, Ultramar, leg. 108, notas, Gobernacion de Ultramar, Sección de Fomento, Negociado de Industria.
17. AGI, Ultramar, leg. 108, December 29, 1821, letter.
18. These lists are reproduced in full in Paloma Blanco et al., "Plantas Cubanas y Documentos de la Ossa en el Real Jardín Botánico de Madrid," *Fontqueria* 36 (1993): 117–46.
19. Mercedes Valero González and Armando García González, "Ciencia y Coleccionismo en Cuba en el Siglo XIX," *Asclepio* 51, no. 1 (1999): 210.
20. Many scholars have written on this history. See especially David Mackay, *In the Wake of Cook: Exploration, Science, and Empire, 1780–1801* (London: Croom Helm, 1985); Richard Drayton, "Science and the European Empires," *Journal of Imperial and Commonwealth History* 23, no. 3 (September 1, 1995): 503–10; John Gascoigne, *Science in the Service of Empire: Joseph Banks, the British State, and the Uses of Science in the Age of Revolution* (Cambridge: Cambridge University Press, 1998); Brockway, *Science and Colonial Expansion.*
21. James E. McClellan III, *Colonialism and Science: Saint Domingue in the Old Regime* (Chicago: University of Chicago Press, 1992), 162.
22. On Suriname, see P. Boomgaard, ed., *Empire and Science in the Making: Dutch Colonial Scholarship in Comparative Global Perspective, 1760–1830,* 1st edition, Palgrave Studies in the History of Science and Technology (New York: Palgrave Macmillan, 2013); on Jamaica, see William Fawcett, *The Public Gardens and Plantations of Jamaica* (Chicago: University of Chicago Press, 1897), 349.
23. Mark Nesbitt, "Botany in Victorian Jamaica," in *Victorian Jamaica,* ed. Tim Barringer and Wayne Modest (Durham: Duke University Press, 2018), 212.

24. Jefferson Dillman, *Colonizing Paradise: Landscape and Empire in the British West Indies* (Tuscaloosa: University of Alabama Press, 2015), 125.
25. On the spread of Linnaean systems, see especially Staffan Muller-Wille, "Collection and Collation: Theory and Practice of Linnean Botany," *Studies in History and Philosophy of Biological and Biomedical Sciences* 38 (2007): 541–62. For more on the political implications of a Linnaean garden in the space of a colonial botanical garden, see especially Antonio Lafuente and Nuria Valverde's essays in Londa L. Schiebinger and Claudia Swan, eds., *Colonial Botany: Science, Commerce, and Politics in the Early Modern World* (Philadelphia: University of Pennsylvania Press, 2005); Antonio Lafuente, "Enlightenment in an Imperial Context: Local Science in the Late-Eighteenth-Century Hispanic World," *Osiris* 15, Nature and Empire: Science and the Colonial Enterprise (2000): 155–73.
26. María Dolores Higueras and José Guío y Sánchez, eds., *Cuba Ilustrada: La Real Comisión de Guantánamo, 1796–1802,* vol. 2 (Barcelona: Lunwerg Editores, 1991), 16.
27. RJB, I.9.4.1, September 11, 1796, letter.
28. *Memorias,* vol. 5 (1818), 294–302, cited in Puig-Samper and Valero, *Historia del Jardín Botánico de la Habana,* 78.
29. *Memorias,* vol. 1 (1817), 23.
30. Communicaciones del Virrey sobre la expedición de Nueva España, Flora Española—1785–98, AMCN, cited in Engstrand, *Spanish Scientists in the New World,* 165.
31. Guadalupe García, *Beyond the Walled City: Colonial Exclusion in Havana* (Berkeley: University of California Press, 2015), 112–13.
32. For more, see especially Matt D. Childs, *The 1812 Aponte Rebellion in Cuba and the Struggle Against Atlantic Slavery* (Chapel Hill: University of North Carolina Press, 2006); Fischer, *Modernity Disavowed.*
33. For the racial politics of the Academy, see especially Fischer, *Modernity Disavowed,* chap. 2.
34. Paul B. Niell, "Rhetorics of Place and Empire in the Fountain Sculpture of 1830s Havana," *The Art Bulletin* 95, no. 3 (September 1, 2013): 440–64.
35. *Memorias,* vol. 26 (1819), 23; Paul B. Niell, *Urban Space as Heritage in Late Colonial Cuba: Classicism and Dissonance on the Plaza de Armas of Havana, 1754–1828* (Austin: University of Texas Press, 2015).
36. AGI, Santo Domingo, leg. 1.714, quoted in Puig-Samper and Valero, *Historia del Jardín Botánico de la Habana,* 96.
37. AGI, Ultramar, leg. 108, num. 10. Also mentioned in Archivo Nacional de Cuba, Intendencia de Hacienda, leg. 318, num. 33, cited in Puig-Samper and Valero, *Historia del Jardín Botánico de la Habana,* 82.
38. ANC, Intendencia de Hacienda, leg. 318, num. 33, cited in Puig-Samper and Valero, *Historia del Jardín Botánico de la Habana,* 82.
39. Provision gardens were common on Cuban plantations, and were even legally required in

a set of regulations on the treatment of enslaved workers passed in 1842. The formerly enslaved man Esteban Montejo described these gardens in his autobiography thus: "These plots of land were the salvation of many slaves, where they got their real nourishment from. Almost all of them had their little strips of land to be sown close to the barracoons [the barrack-like buildings where the enslaved workers slept], almost behind them. Everything grew there: sweet potatoes, gourds, okra, kidney beans, which were like lime beans, yucca and peanuts. They also raised pigs. And they sold all these products to the whites who came out from their villages." Esteban Montejo, *The Autobiography of a Runaway Slave,* ed. Miguel Barnet, trans. Jocasta Innes (London: Bodley Head, 1966), 24. Provision gardens of enslaved peoples served not just as alimentary necessity, however. As suggested by Montejo above, enslaved people were often able to sell surplus crops from their gardens for cash, which was often used to buy their own freedom or the freedom of loved ones.

40. Katherine McKittrick and Clyde Adrian Woods, eds., *Black Geographies and the Politics of Place* (Toronto: Between the Lines and South End Press, 2007).
41. Louis Nelson and Maurie McInnis, "Landscape of Slavery," in *Educated in Tyranny: Slavery at Thomas Jefferson's University* (London: University of Virginia Press, 2019), 43.
42. Upton, "White and Black Landscapes in Eighteenth-Century Virginia," 66.
43. Judith Ann Carney and Richard Nicholas Rosomoff, *In the Shadow of Slavery: Africa's Botanical Legacy in the Atlantic World* (Berkeley: University of California Press, 2011), 123.
44. *Anales de Ciencias, Agricultura, Comercio y Artes,* vol. 1 (Havana: Oficina del Gobierno y Capitanía General, 1827).
45. ANC, leg. 1124, num. 72, March 17, 1828, letter from Sagra.
46. AGI, Ultramar, leg. 108, num. 8.
47. Camilla Townsend, "'The More I See, the More Surprised I Am': Ramón de La Sagra, Baltimore, and the Concepts of Race of Poverty," in *Strange Pilgrimages: Exile, Travel, and National Identity in Latin America, 1800–1990s,* ed. Ingrid Elizabeth Fey and Karen Racine (Wilmington: Scholarly Resources, 2000), 41–58, is a particularly astute summary of Sagra's early life and complex feelings about Cuba.
48. Archivos Fondos Real Jardín Botánico, I.32.5.4.
49. *Memorias,* vol. 5 (1825), 790.
50. AGI, Ultramar, leg. 149.2, Dec. 3, 1827, letter from Don Jacobo María de Parga; emphasis added.
51. Translation from Francisco de Arango y Parreño, "Discurso sobre la Agricultura de la Habana y Medios de Fomentarla," in *Obras* (Havana: Dirección de Cultura, Ministerio de Educación, 1952), 114–75; translation from Dale Tomich, "The Wealth of Empire: Francisco Arango y Parreño, Political Economy, and the Second Slavery in Cuba," *Comparative Studies in Society and History* 45, no. 1 (2003): 8.

52. Manuel Moreno Fraginals, *The Sugarmill: The Socioeconomic Complex of Sugar in Cuba, 1760–1860,* trans. Cedric Belfrage (New York: Monthly Review, 1976), 19.
53. Puig-Samper and Valero, *Historia del Jardín Botánico de la Habana,* 29.
54. *Memorias,* vol. 1 (1793), 147–48, cited in Puig-Samper and Valero, 30–31.
55. "Memorias of the Sociedad Patriótica of Havana in the year 1794," reprinted in *Memorias de la Sociedad Económica de la Habana,* vol. 17 (1843), 420.
56. *Memorias,* vol. 5 (1825), 785.
57. Ramón de la Sagra, *Informe Sobre el Estado Actual del Jardín y de la Cátedra de Botánica Applicada a la Agricultura* (Havana: Sociedad Patriótica de la Habana, 1825), 4.
58. Sagra, *Informe Sobre el Estado Actual,* 12.
59. Sagra, *Informe Sobre el Estado Actual,* 7.
60. *Memorias,* vol. 5 (1825), 785, 788.
61. Chira, *Patchwork Freedoms,* 118.
62. This *cédula* specified that "white" in this context meant white Spaniards from the Peninsula or the Canary Islands, or Catholics from European countries who were allies of Spain and granted Spanish nationality to any such settlers who remained in Cuba for five or more years. In practice, most white immigrants in this policy were French, many of them sugar mill owners from former Saint Domingue. *Real Cédula de 21 de Octubre de 1817, Sobre Aumentar la Población Blanca de la Isla de Cuba* (Imprenta del Gobierno y Capitanía General, 1828), 2. See also "Expediente Formado en Virtud de Circular de Alejandro Ramirez, Intendente General, Fecha Habana 5 de Abril de 1819, Sobre Promover la Agricultura y Población Blanca," 1819, leg. 1676, num. 83955, ANC, Gobierno Superior Civil.
63. For more on the Puerto Rico ruling, see especially Ivette Pérez Vega, "Effects of the Spanish Rotal Decree 'Cédula De Gracias' of 1815 on Inmigration in Puerto Rico, 1819–1930," *Revista Umbral* 14 (December 2018): 113–40.
64. Niell, *Urban Space as Heritage in Late Colonial Cuba,* 71.
65. Kay Dian Kriz, *Slavery, Sugar, and the Culture of Refinement: Picturing the British West Indies, 1700–1840* (New Haven: Yale University Press, 2008), 158.
66. Ernesto Bassi, "The 'Franklins of Colombia': Immigration Schemes and Hemispheric Solidarity in the Making of a Civilised Colombian Nation—ProQuest," *Journal of Latin American Studies* 50, no. 3 (August 2018): 673–701; Erika Denise Edwards, *Hiding in Plain Sight: Black Women, the Law, and the Making of a White Argentine Republic* (Tuscaloosa: University of Alabama Press, 2020).
67. Chira, *Patchwork Freedoms,* 118.
68. ANC, leg. 1676, num. 83955, "Agricultura y población blanca."
69. *Memorias de la Real Sociedad Patriótica de la Habana,* vol. 1 (1836), 25–26.
70. Fischer, *Modernity Disavowed.*

71. Ada Ferrer, *Freedom's Mirror: Cuba and Haiti in the Age of Revolution* (Cambridge: Cambridge University Press, 2014). For more on this, see also Chira, *Patchwork Freedoms;* Manuel Barcia Paz, *The Great African Slave Revolt of 1825: Cuba and the Fight for Freedom in Matanzas* (Baton Rouge: Louisiana State University Press, 2012).
72. For more on this migration, see especially Shalini Puri, ed., *Marginal Migrations: The Circulation of Cultures Within the Caribbean,* Warwick University Caribbean Studies (Oxford: Macmillan Caribbean, 2003); Margarita Cervantes-Rodriguez and Alejandro Portes, *International Migration in Cuba: Accumulation, Imperial Designs, and Transnational Social Fields* (University Park: Pennsylvania State University Press, 2011); Naranjo Orovio and García González, *Racismo e Inmigración en Cuba en el Siglo XIX.*
73. Thomas F. Anderson, *Carnival and National Identity in the Poetry of Afrocubanismo* (Gainesville: University Press of Florida, 2011); Sartorius, *Ever Faithful;* Daniel E. Walker, *No More, No More: Slavery and Cultural Resistance in Havana and New Orleans* (Minneapolis: University of Minnesota Press, 2004).
74. Archivo General de Indias, Ultramar, leg. 149.2, Sept. 28, 1830, letter from palace; Sept. 27, 1833, letter from Intendente of Havana.
75. AHNE, Ultramar, leg. 2, num. 25.
76. Ana Amigo, "Identidad, Modernidad, Ocio: Jardines Urbanos de la Habana en el Siglo XIX," *Cuban Studies* 46 (2018): 87–112.
77. See, for instance, *Memorias de la Real Sociedad Económica de la Habana,* vol. 5 (1825), 597 and 770.
78. *Memorias,* vol. 5 (1825), "Esposicion de las tareas que han ocupado a la Real Sociedad Patriótica durante los anos de 1825 y 1826, leida en junta general de 15 de diciembre," D. Joaquín Santos Suárez, 30.
79. Sagra, *Informe Sobre el Estado Actual del Jardín y de la Cátedra de Botánica Applicada a la Agricultura,* 5.
80. Townsend, "'The More I See, the More Surprised I Am,'" 45.
81. *Memorias,* vol. 5 (1825), 607.
82. For more on these debates, see especially Gloria García Rodríguez, Nancy L. Westrate, and Ada Ferrer, *Voices of the Enslaved in Nineteenth-Century Cuba: A Documentary History,* Latin America in Translation/En Traducción/Em Tradução (Chapel Hill: University of North Carolina Press, 2011).
83. *Diario de la Habana* (Havana), May 3, 1839, 1–2.
84. Joseph John Gurney, *A Winter in the West Indies Described in Familiar Letters to Henry Clay, of Kentucky,* electronic resource (London: J. Murray, 1840), 210.
85. James M. Phillippo, *The United States and Cuba* (London: Pewtress; Sheldon, Blakeman, 1857), 563.
86. Fanny Elssler, *The Letters and Journal of Fanny Ellsler: Written Before and After Her Operatic*

Campaign in the United States; Including Her Letters from New York, London, Paris, Havana (New York: H.G. Daggers, 1845), 44.

87. Samuel Hazard, *Cuba with Pen and Pencil* (London: Sampson, Low, Marston, Low & Searle, 1873), 139.
88. Ian McLean, "The Expanded Field of the Picturesque: Contested Identities and Empire in Sydney-Cove 1794," in *Art and the British Empire,* ed. Geoff Quilley, Douglas Fordham, and Tim Barringer (Manchester: Manchester University Press, 2007), 26.
89. Archivo Histórico Nacional de España, Ultramar, leg. 226, exp. 7, letters from August 27, 1865, and December 18, 1865.

2. *"The Peculiar Circumstances of Cuba"*

1. García, *Beyond the Walled City.*
2. Niell, *Urban Space as Heritage in Late Colonial Cuba.*
3. Niell, *Urban Space as Heritage,* 147.
4. BNJM, CM Sociedad, t. 46, num. 7.
5. *Memorias de la Sociedad Patriótica de la Habana,* vol. 11 (1840), 201.
6. Felipe Poey, *Centurie de Lépidoptères de l'Ile de Cuba* (Paris: J. Albert Mercklein, 1832); Felipe Poey, *Compendio de la Geografía de la Isla de Cuba* (Havana: Imprenta del Gobierno y Capitanía General, 1836).
7. Dr. Julian Vivanco y Diaz, *Don Felipe: Su Vida y Su Obra* (Havana: Academia de la Historia de Cuba y Imprenta "El Siglo XX," 1951).
8. *Memorias de la Real Sociedad Patriótica de la Habana,* vol. 6 (1838), 25.
9. *Memorias,* vol. 11 (1840), 210.
10. *Memorias,* vol. 11 (1840), 203.
11. *Memorias,* vol. 11 (1840), 207.
12. For more on the centrality of Paris, see, for instance, E. C. Spary, *Utopia's Garden: French Natural History from Old Regime to Revolution* (Chicago: University of Chicago Press, 2000).
13. *Memorias,* vol. 11 (1840), 203.
14. *Memorias,* vol. 11 (1840), 203, and *Memorias,* vol. 6 (1838), 25.
15. For more on this moment, see especially Sartorius, *Ever Faithful.*
16. *Memorias de la Real Sociedad Económica de la Habana,* vol. 5 (1825), 375; emphasis added.
17. Vivanco y Diaz, *Don Felipe: Su Vida y Su Obra,* 16.
18. Letter from Felipe Poey to José Chauvrieau, Madrid, March 4, 1822, archives of Alejandro Barro, University of Havana.
19. Felipe Poey, *Obras,* ed. Rosa María González López (Havana: Imagen Contemporánea, 1999), 327.

20. *Memorias de la Sociedad Patriótica de la Habana,* vol. 14 (1842), 158.
21. *Memorias,* vol. 6 (1838), 21.
22. *Memorias,* vol. 11 (1840), 206.
23. *Memorias de la Real Sociedad Patriótica de la Habana,* vol. 6 (1838), 27.
24. *Memorias,* vol. 7 (1838), 176.
25. *Memorias,* vol. 11 (1840), 202.
26. *Diario de la Habana,* June 20, 1838, 2.
27. Dorinda Outram, "New Spaces in Natural History," in *Cultures of Natural History,* ed. N. Jardine, J. A. Secord, and E. C. Spary (Cambridge: Cambridge University Press, 1996), 250. For more on the shift in scientific knowledge production in the mid-nineteenth century, see especially Daston and Galison, *Objectivity.*
28. Andrew J. Lewis, *A Democracy of Facts: Natural History in the Early Republic* (Philadelphia: University of Pennsylvania Press, 2011), 15.
29. *Memorias de la Sociedad Patriótica de la Habana,* vol. 10 (1840), 85; *Memorias de la Real Sociedad Patriótica de la Habana,* vol. 9 (1839), 9.
30. José Luis Alfanso, the nobleman living in Paris, donated a table made with 164 different types of marble; see *Memorias,* vol. 8 (1839), 3. Francisco Ruiz [dates unknown] donated a set of 18 mineral specimens that he had discovered in the Vuelta-Abajo region. Ramón de la Paz y Morejón, a priest in Yaguaramas, sent three shipments of unnamed objects; *Memorias,* vol. 9 (1839), 6. Manuel Valdes Miranda, a doctor, donated a small collection of conches, and Tomás Juara y Soler, a sugar mill owner, gave the museum a canoe made by indigenous Canadians; see *Memorias,* vol. 7 (1838), 323; *Memorias,* vol. 11 (1840), 75.
31. Philip Henry Gosse and Richard Hill, *A Naturalist's Sojourn in Jamaica* (London: Longman, Brown, Green and Longmans, 1851); Michel Etienne Descourtilz, *Flore Pittoresque et Médicale des Antilles, Ou, Histoire Naturelle des Plantes Usuelles des Colonies Françaises, Anglaises, Espagnoles et Portugaises* (Paris: Pichard, 1821).
32. García, *Beyond the Walled City,* 111, 113.
33. For more on Peale's museum, see especially David C. Ward, *Charles Willson Peale: Art and Selfhood in the Early Republic* (Berkeley: University of California Press, 2004); Susan Stewart, "Death and Life, in That Order, in the Works of Charles Willson Peale," in *The Open Studio: Essays on Art and Aesthetics* (Chicago: Chicago University Press, 2005), 185–202; Laura Rigal, "Peale's Mammoth," in *American Iconology: New Approaches to Nineteenth-Century Art and Literature,* ed. David C. Miller (New Haven: Yale University Press, 1993), 18–38; David R. Brigham, "Social Class and Participation in Peale's Philadelphia Museum," in *Mermaids, Mummies, and Mastadons: The Emergence of the American Museum,* ed. William T. Alderson (Washington, D.C.: The American Association of Museums, 1992), 79–88.

34. *Memorias de la Sociedad Económica de la Habana,* vol. 20 (1845), 234.
35. *Memorias de la Real Sociedad Económica de la Habana,* vol. 7.2 (1849), 4.
36. *Memorias,* vol. 7.2 (1849), 4.
37. *Memorias de la Sociedad Económica de la Habana,* vol. 18 (1844), 297.
38. *Memorias,* vol. 18 (1844), 306.
39. On agricultural fairs, see Wayne Neely, *The Agricultural Fair* (New York: Columbia University Press, 1935); Guy Scott, *Country Fairs in Canada* (Markham: Fitzhenry & Whiteside, 2005); Leslie Prosterman, *Ordinary Life, Festival Days: Aesthetics in the Midwestern County Fair* (Washington, D.C.: Smithsonian Institution Press, 1995). On their relation to the British country fair, see Dmytro Ostapenko, "Generating New Production Knowledge: Competitive Agricultural Events in the British Australian Colony of Victoria, c. 1840–1890," *Agricultural History* 92, no. 4 (2018): 491–511; Emily Pawley, "Cataloging Nature: Standardizing Fruit Varieties in the United States, 1800–1860," *Business History Review* 90, no. 3 (Autumn 2016): 405–29.
40. See, for instance, "Record of News," *Friends' Weekly Intelligencer,* September 20, 1845.
41. *Memorias,* vol. 18 (1844), 293.
42. *Memorias,* vol. 18 (1844), 255.
43. *Memorias,* vol. 18 (1844), see especially page 299.
44. *Memorias,* vol. 18 (1844), see especially page 299.
45. *Memorias,* vol. 18 (1844), 307.
46. *Memorias,* vol. 2 (1836), 35, for instance.
47. *Memorias,* vol. 18 (1844), 309.
48. On this period, see especially Robert L. Paquette, *Sugar Is Made with Blood: The Conspiracy of La Escalera and the Conflict Between Empires over Slavery in Cuba* (Middletown: Wesleyan University Press, 1988), Michele Reid-Vazquez, *The Year of the Lash;* Sara E. Johnson, *The Fear of French Negroes: Transcolonial Collaboration in the Revolutionary Americas* (Berkeley: University of California Press, 2012), Jonathan Curry-Machado, "How Cuba Burned with the Ghosts of British Slavery: Race, Abolition, and the Escalera," *Slavery & Abolition* 25, no. 1 (April 1, 2004): 71–93, Daniel E. Walker, *No More, No More,* and Finch, *Rethinking Slave Rebellion.*
49. Louis A. Pérez, *Winds of Change: Hurricanes and the Transformation of Nineteenth Century Cuba* (Chapel Hill: University of North Carolina Press, 2001).
50. See especially the introduction of Paquette, *Sugar Is Made with Blood.*
51. Paquette, *Sugar Is Made with Blood,* 4.
52. For more, see especially the introduction of Reid-Vazquez, *The Year of the Lash.*
53. ANC, Gobierno Superior Civil, leg. 1603, num. 81764. It is likely, of course, that this delay was also due to the upheaval of the Year of the Lash, though this is not mentioned.

54. *Memorias,* vol. 5.2 (1848), 139–40.
55. *Memorias,* vol. 5.2 (1848), 167.
56. *Memorias,* vol. 5.2 (1848), 167.
57. For more, see especially Fischer, *Modernity Disavowed,* ch. 2.
58. Junta Ordinaria, November 17, 1817, in *Acuerdos de la Junta Ordinaria de la Real Sociedad Patriótica,* book 6, MS sheets 67–68; quoted in and translated by Fischer, *Modernity Disavowed,* 74.
59. José Antonio Saco López, *Memoria Sobre la Vagancia en la Isla de Cuba* (Santiago de Cuba: Instituto Cubano del Libro, 1874), 58–59, quoted in and translated by Fischer, *Modernity Disavowed,* 73.
60. Eduardo Torres-Cuevas, Jorge Ibarra, and Mercedes García, eds., *Obras de Félix Varela* (Havana: Editora Política, 1991), 116, cited in Michele Reid-Vazquez, *The Year of the Lash,* 31.
61. Reid-Vazquez, *The Year of the Lash,* 104, 29.
62. Reid-Vazquez, *The Year of the Lash,* 104.
63. For more on economic policy during this period, see especially Louis A. Pérez Jr., *Cuba Between Reform and Revolution* (Oxford: Oxford University Press, 2010).
64. Sagra himself visited the Exposition and wrote an account of it that was published in Madrid in 1853, which focused on the agricultural technology on display there. Ramón de la Sagra, *Memoria Sobre los Objetos Estudiados en la Exposition Universal de Lóndres y Fuera de Ella, Bajo el Punto de Vista del Adelanto Futuro de La Agricultura é Industria Españoles* (Madrid: Imprenta del Ministerio de Fomento, 1853).
65. BNJM, CM Sociedad, tome 28, num. 6, object 17.
66. Vicente Casals Costa, "Las Ideas Sobre la Protección del Bosque en Cuba y Filipinas Durante el Siglo XIX," in *Ciencia, Vida y Espacio en Iberoamérica,* ed. José Luis Peset Reig, vol. 3, Estudios Sobre la Ciencia 10 (Madrid: CSIC, 1989). See also Reinaldo Funes Monzote, *From Rainforest to Cane Field in Cuba: An Environmental History Since 1492* (Chapel Hill: University of North Carolina Press, 2008).
67. Manuel Moreno Fraginals and Luis M. de las Traviesas Moreno, *El Ingenio: Complejo Económico Social Cubano del Azúcar, Nuestra Historia* (Havana: Editorial de Ciencias Sociales, 1978), 158.
68. Monzote, *From Rainforest to Cane Field in Cuba,* 157.
69. Natalis Rondot, *Catalogue Officiel, Publié par Ordre de la Commission Impériale* (Paris: E. Panis, 1855), 484–85.
70. *International Exhibition, 1876: Official Catalogue,* Part 1, Main building and annexes; Department 1—Mining and metallurgy; Department 2—Manufactures; Department 3—Education and science (Philadelphia: John R. Nagle, 1876), 270–79.
71. BNJM, Exposición de Amsterdam, CM Sociedad, tome. 14, num. 13.

3. "The Novelty of the Productions of the Island of Cuba"

1. Antonio Parra, *Descripción de Diferentes Piezas de Historia Natural, las Mas del Ramo Marítimo, Representadas en Setenta y Cinco Laminas* (Havana, 1787).
2. Theodor de Bry, *Collectiones peregrinationum in Indiam Occidentalem et Indiam Orientalem* (Francoforti at Moenvm; Beckerum: Thoedore de Bry: 1590).
3. Juan Lembeye, *Aves de La Isla de Cuba* (Havana: Imprenta del Tiempo, 1850).
4. These were the botanical paintings created by José Guío y Sánchez for the Real Comisión de Guantánamo but now known primarily as the Mopox expedition, and those created by Juan de Dios Vicente de la Cerda and Atanasio Echeverría y Godoy for the Cuban section of the Real Expedición Botánica a Nueva España, now known primarily as the Sessé expedition. For more on these paintings, see especially Higueras and Guío y Sánchez, *Cuba Ilustrada;* Miguel Ángel Puig-Samper Mulero and María Pilar de San Pío Aladrén, eds., *Las Flores del Paraíso: La Expedición Botánica de Cuba en los Siglos XVIII y XIX* (Barcelona: Real Jardin Botánico CSIC, 1999); Puig-Samper, "Illustrators of the New World: The Image in the Spanish Scientific Expeditions of the Enlightenment."
5. Juan Sanchez, *El Grabado en Cuba* (Havana: Impresora Mundial, 1955); Zoila Lapique Becali, *La Memoria en las Piedras* (Havana: La Oficina del Historiador de la Ciudad, 2002); On natural history illustration specifically, see Emilio C. Cueto, *Illustrating Cuba's Flora and Fauna* (Miami: Historical Museum of Southern Florida, 2002).
6. Ramón de la Sagra, *Historia Física, Política y Natural de la Isla de Cuba* (Paris: En libreria de Arthus Bertrand, 1838).
7. Felipe Poey, *Memorias Sobre la Historia Natural de la Isla de Cuba, Acompañadas de Sumarios Latinos y Exractos en Frances* (Havana: Imprenta de Barcina, 1851); Felipe Poey, *Memorias Sobre la Historia Natural de la Isla de Cuba, Acomañadas de Sumarios Latinos y Exractos en Frances,* vol. 2 (Havana: Imprenta de Viuda de Barcina, 1856).
8. AGI, Ultramar, 149.2, Sept. 23, 1834, letter from Ministerio de Interior.
9. BNJM, CM Escoto, num. 138, March 27, 1835, letter from Sagra.
10. *Prospecto del Historia Física, Política y Natural de la Isla de Cuba* (Paris: F.G. Levrault, 1837), AGI, Ultramar, 149.2.
11. Mimi Sheller, *Consuming the Caribbean: From Arawaks to Zombies* (London: Routledge, 2003), 37. See also, especially, Felix Driver and Luciana Martins, eds., *Tropical Visions in an Age of Empire* (Chicago: University of Chicago Press, 2005); Nancy Leys Stepan, *Picturing Tropical Nature* (Ithaca: Cornell University Press, 2001); Krista S. Thompson, *An Eye for the Tropics: Tourism, Photography, and Framing the Caribbean Picturesque* (Durham: Duke University Press, 2006); Mary Louise Pratt, *Imperial Eyes: Travel Writing and Transculturation,* second edition (London: Routledge, 2008).
12. Driver and Martins, *Tropical Visions in an Age of Empire.*
13. Humboldt, *Political Essay on the Island of Cuba: A Critical Edition,* 23.

14. Alexander von Humboldt, *Views of Nature: Or Contemplations on the Sublime Phenomena of Creation. With Scientific Illustrations* (G. Bell, 1850), 230.
15. Townsend, "'The More I See, the More Surprised I Am.'"
16. BNJM, Expediente Personal, 63.14, March 25, 1836, letter.
17. Beth Fowkes Tobin, "Imperial Designs: Botanical Illustrations and the British Botanic Empire," in *Picturing Imperial Power: Colonial Subjects in Eighteenth-Century British Painting* (Durham: Duke University Press, 1999), 174–201, 179.
18. Alcide Dessalines d'Orbigny, *Voyage Dans l'Amérique Méridionale* (Paris: Chez Pitois-Levrault, 1835–1847).
19. Georges-Louis Leclerc Buffon, Georges Cuvier, and Edouard Traviès, *Oeuvres complètes de Buffon* (Paris: Dufour, Mulat et Boulanger, 1856–1857).
20. Maria Sibylla Merian, *Metamorphosis insectorum Surinamensium: In qua erucae ac vermes Surinamenses ... ad vivum delineantur & describuntur ...* (Amsterdam: Venduntur & apud Gerardum Valk op den Dam, 1705); Mark Catesby and George Edwards, *The Natural History of Carolina, Florida, and the Bahama Islands* (London: B. White, 1771).
21. The most famous of these was Alcide d'Orbigny, who wrote the fourth volume, on birds, as well as the fifth volume, on mollusks. Other authors Sagra worked with included Jean Théodore Cocteau, Gabriel Bibron, Alphone Guichenot, Camille Montagne, and Félix Édouard Guérin-Méneville.
22. Sagra, *Histoire Physique, Politique et Naturelle de l'ile de Cuba : Reptiles,* by J. T. Cocteau and Gabriel Bibron, 61–62.
23. Ramón de la Sagra and M. Alcide D'Orbigny, *Histoire Physique, Politique et Naturelle de l'ile de Cuba,* trans. M. S. Berthelot, vol. 3: Ornitologie (Paris: A. Bertrand, 1838), viii; emphasis added.
24. Sagra and D'Orbigny, *Histoire Physique, Politique et Naturelle,* vol. 3: 24.
25. Ramón de la Sagra and Camille Montagne, *Histoire Physique, Politique et Naturelle de l'ile de Cuba,* vol. 9: Botanique—Plantes Cellulaires (Paris: A. Bertrand, 1838), 27.
26. In thinking about natural history information being refined in the metropole, just as sugar is refined, I look to Kriz, *Slavery, Sugar, and the Culture of Refinement: Picturing the British West Indies, 1700–1840.*
27. Ramón de la Sagra, *Histoire Physique, Politique et Naturelle de l'Ile de Cuba,* trans. M. S. Berthelot, vol. 2: Agriculture, commerce, revenus et appendice (Paris: A. Bertrand, 1845), 47–48.
28. Sagra, *Histoire Physique, Politique et Naturelle,* vol. 2: 36, 29.
29. A sense of this is provided in the reports from Madrid on which sets the court had received at different points of time. These are held in AGI, ULTRAMAR, 149.2. See, for instance, the list dated May 6, 1843.
30. For more on this, see especially Paquette, *Sugar Is Made with Blood,* 37.

31. Ramón de la Sagra, *Anales de Ciencias, Agricultura, Comercio y Artes,* vol. 2 (Havana: Oficina del Gobierno y Capitanía General, 1828), 25.
32. I have consulted the Database of Scientific Illustrators, located at www.uni-stuttgart.de/hi/gnt/dsi2/, to confirm full names and dates for the illustrators in Sagra's *Historia.*
33. Fr.-Richard de Tussac, *Flore des Antilles, ou Histoire générale, botanique, rurale et économique des végétaux indigènes des Antilles, et des exotiques qu'on est parvenu à y naturaliser* (Paris: L'auteur, 1808); Descourtilz, *Flore Pittoresque et Médicale des Antilles.*
34. BNJM, CM Socieded T17 num. 18 a-e, "Comunicación de D. Ramón de la Sagra, 1824–1826," December 18, 1826, letter from Sagra.
35. Daston and Galison, *Objectivity,* chap. 1.
36. BNJM, Expediente Personal, 63.14, February 16, 1836, letter.
37. AGI, Ultramar, 149.2, November 1, 1841, letter from Havana.
38. AGI, Ultramar, 149.2.
39. AGI, Ultramar, 149.2, printed pamphlet dated June 27, 1854.
40. MNCN, ACN, 0400.018.
41. Sagra, *Histoire,* vol. 1: xii.
42. BNJM, Expediente Personal, 63.14, Janunary 8, 1835, letter.
43. Sujit Sivasundaram, "Islanded: Natural History in the British Colonization of Ceylon," in *Geographies of Nineteenth-Century Science,* ed. David N. Livingstone and Charles W. J. Withers (Chicago: University of Chicago Press, 2011), 143, speaks eloquently of the "fragility of Kew's reach" at this moment.
44. Poey, *Memorias Sobre la Historia Natural de la Isla de Cuba,* vol. 1: 3.
45. Darío Guitart Manday, "Nota Introductoria," in Felipe Poey y Aloy, *Ictiología Cubana,* vol. 1, Biblioteca de Clásicos Cubanos (Havana: Centro de Investigaciones Marinas de la Universidad de la Habana y el Ministerio de la Industria Pesquera, Ediciones Imagen Contemporánea, 2000), unpaginated.
46. Poey, *Memorias Sobre la Historia Natural de la Isla de Cuba,* vol. 1: 16, 4.
47. "This accredited artist, who knows that common wax enters 40 percent in the composition of lithographic ink, according to Engelmann's recipe, which does not contain tallow, had the happy idea of composing a new ink, creating a new recipe and replacing the white wax with brown wax. The results were extremely satisfactory." Poey, *Memorias Sobre la Historia Natural de la Isla de Cuba,* vol. 1: 169. See also a discussion of what position to draw mollusks in on page 16 of volume 1.
48. Sociedad Española de Historia Natural, *Anales de la Sociedad Española de Historia Natural,* vol. 1 (Madrid: La Sociedad, 1872), 231; Bernard Quaritch, *Catalogue of Works on Natural History, Physics, Mathematics, and Other Sciences,* vol. 2 (G. Norman and Son, 1881), 292.
49. Poey, *Memorias Sobre la Historia Natural de la Isla de Cuba,* vol. 1: 3.

50. *Revue Zoologique, par la Société Cuvierienne,* editor M. F.-E. Guérin-Méneville (Paris, 1841), 269.
51. Poey, *Memorias Sobre la Historia Natural de la Isla de Cuba,* vol. 1: 29.
52. Poey, *Memorias Sobre la Historia Natural de la Isla de Cuba,* vol. 1: 380, 206.
53. Poey, *Memorias Sobre la Historia Natural de la Isla de Cuba,* vol. 1: 53.
54. This is especially true in chapter 10 in the first volume of *Memorias Sobre la Historia Natural,* in which Poey notes that certain species were "found in Triunidad by Sr,. Lavallee," 110, "found in Baracoa by Sres. Paz and Rios," 112, "brought from Trinidad by D. Julio Sagebien," 115, and "encountered in Cayajabos by D. Nazareno de la Paz," 116.
55. Poey, *Memorias Sobre la Historia Natural de la Isla de Cuba,* vol. 2: 351.
56. Poey, *Memorias Sobre la Historia Natural de la Isla de Cuba,* vol. 2: 17–23.
57. Joseph Dalton Hooker et al., *Himalayan Journals, or, Notes of a Naturalist in Bengal, the Sikkim and Nepal Himalayas, the Khasia Mountains, &c* (London: John Murray, 1854). For more, see especially Mary Louise Pratt, *Imperial Eyes: Travel Writing and Transculturation* (London: Routledge, 1992); Gillian Beer, "Travelling the Other Way: Travel Narratives and Truth Claims," in *Open Fields: Science in Cultural Encounter* (Oxford: Clarendon Press, 1996), 55–70.
58. *Memorias,* vol. 7.2 (1849), 137.
59. BNJM, CM Bachiller, num. 343.
60. David S. Jordan, "Sketch of Professor Felipe Poey," *Popular Science Monthly* (August 1884), 548.
61. Felipe Poey y Aloy, *Ictiología Cubana,* ed. Mario Sánchez Roig and Federico Gómez de la Maza, vol. 1 (Havana: República de Cuba Ministerio de Educación, 1955), 70.
62. Poey, *Memorias Sobre la Historia Natural de la Isla de Cuba,* vol. 1: 27.
63. Poey, *Memorias Sobre la Historia Natural de la Isla de Cuba,* vol. 1: 58, 103, and 208, among others.
64. Poey, *Memorias Sobre la Historia Natural de la Isla de Cuba,* vol. 1: 4.
65. Poey, *Memorias Sobre la Historia Natural de la Isla de Cuba,* vol. 1: 115.
66. For instance, see Londa L. Schiebinger and Claudia Swan, eds., *Colonial Botany: Science, Commerce, and Politics in the Early Modern World* (Philadelphia: University of Pennsylvania Press, 2005), and Stephan Palmié, *Wizards and Scientists: Explorations in Afro-Cuban Modernity and Tradition* (Durham: Duke University Press, 2002).
67. Monzote, *From Rainforest to Cane Field in Cuba;* see also Jonathan Curry-Machado, *Transnational Networks and Engineering Migrants in Mid-Nineteenth Century Cuba* (New York: Palgrave Macmillan, 2011).
68. Arthur Morelet, "Observaciones sobre la conservación de arbolados: Carta remitida por Arturo Moselet [*sic*] en su viaje por la isla en al año 1847," *Memorias de la Real Sociedad*

Económica de la Habana, vol. 37 (1848), 93; cited in Monzote, *From Rainforest to Cane Field in Cuba,* 158.

69. Monzote, *From Rainforest to Cane Field in Cuba,* 149–52.
70. Vicente Casals Costa, "Las Ideas Sobre la Protección del Bosque en Cuba y Filipinas Durante el Siglo XIX," in *Ciencia, Vida y Espacio en Iberoamérica,* ed. José Luis Peset Reig, vol. 3, Estudios Sobre la Ciencia 10 (Madrid: CSIC, 1989), 373.
71. Poey, *Memorias Sobre la Historia Natural de la Isla de Cuba,* vol. 1: 86, 78.
72. Poey's representation of Cuban honeybees focuses on the dense relationships between their bodies, the parts of their anatomy, and the hive structures that they live in, which is quite different from, for instance, Joris Hoefnagel's illustration of a bee from his Four Elements manuscripts. Marisa Bass discusses how in this illustration, produced in the late sixteenth-century Low Countries, Hoefnagel uses the insect to think through the relationships between the independent individual and the collective society, as well as the duality of human and animal nature. Marisa Anne Bass, *Insect Artifice: Nature and Art in the Dutch Revolt* (Princeton: Princeton University Press, 2019), 4.
73. José Alvarez Conde, *Historia de la Zoología en Cuba* (Havana: Publicaciones de la Junta Nacional de Arqueología y Etnología, 1958), provides a full list of these varous publications—I focus specifically on the illustrated texts from this period.
74. Lembeye, *Aves de la Isla de Cuba,* 6–7.
75. Lembeye, *Aves de la Isla de Cuba,* 7.
76. Cueto, *Illustrating Cuba's Flora and Fauna,* 59.
77. A. Grisebach and Charles Wright, *Plantae Wrightianae e Cuba Orientali* (Canteabriglae Nov. Angl., 1860); A. Grisebach, *Catalogus Plantarum Cubensium* (Lipsiae: Apud Guilielmum Engelmann, 1866).
78. Francisco Sauvalle, *Flora Cubana* (Havana: La Antilla, 1873).
79. Richard A. Howard, *Charles Wright in Cuba, 1856–1867* (Alexandria: Chadwyck-Healey, 1988).
80. Howard, *Charles Wright in Cuba,* 39.
81. Jennifer L. Roberts, *Transporting Visions: The Movement of Images in Early America* (Berkeley: University of California Press, 2014), 2.
82. These partial publications have included: Felipe Poey, *Ictiología Cubana,* ed. Pedro Pablo Duarte Bello (Havana: Comisión Nacional de la Academía de Ciencias de la República de Cuba, 1962); Felipe Poey y Aloy, *Ictiología Cubana,* ed. Mario Sánchez Roig and Federico Gómez de la Maza (Havana: República de Cuba Ministerio de Educación, 1955); and Felipe Poey y Aloy, *Ictiología Cubana,* Biblioteca de Clásicos Cubanos (Havana: Centro de Investigaciones Marinas de la Universidad de la Habana y el Ministerio de la Industria Pesquera, Ediciones Imagen Contemporánea, 2000).

83. Aldemaro Romero Jr., "Felipe Poey, Hyperbole, and the Myth of the 'Isolated Genius' Among Spanish and Latin American Scientists," *Polymath: An Interdisciplinary Journal of Arts & Sciences* 4, no. 4 (Fall 2014): 85–95.
84. Poey y Aloy, *Ictiología Cubana* (2000).
85. See, for instance, MNCN, ACN 0406.007.

4. "Make Known the Island of Cuba"

1. Described in "Viaje a los Cayos" in Felipe Poey, *Obras literarias de Felipe Poey* (Havana: La Propaganda literaria, 1888).
2. Lapique Becali, *La Memoria en las Piedras,* is the exception to this, with a careful and insightful tracing of the development of Cuban lithography and the various presses during this period. This chapter is indebted to her in-depth examinations of this field.
3. J. B. Harley, "Maps, Knowledge, and Power," in *The Iconography of Landscape: Essays on the Symbolic Representation, Design and Use of Past Environments,* ed. Denis Cosgrove and Stephen Daniels (Cambridge: Cambridge University Press, 1988), 279.
4. Jeffrey Auerbach, "The Picturesque and the Homogenisation of Empire," *The British Art Journal* 5, no. 1 (Summer 2004): 48.
5. Lapique Becali, *La Memoria en las Piedras.*
6. *Diario de la Habana,* August 11, 1837.
7. "La Habana que con tanta rapidez progresa hacia su engrandecimiento, y que hoy marcha á la par de las ciudades más florecientes de América, y aun de Europa carece sin embargo de uno de los elementos indispensables en todo país adelantado en civilización, nos contraemos al establecimiento de una imprenta Litográfica"; *Diario de la Habana,* August 11, 1837.
8. Cueto, *Mialhe's Colonial Cuba,* note 6, p. 8.
9. Lapique Becali, *La Memoria en las Piedras,* 67 and 71.
10. "nos dará los mejores modelos de paisajes Cubanos"; *Diario de la Habana,* June 24, 1839.
11. For example, see Irene Aloha Wright, *Cuba* (New York: Macmillan, 1910), 258.
12. Cueto, *Mialhe's Colonial Cuba,* 9.
13. Lapique Becali, *La Memoria en las Piedras,* 116.
14. Asiel Sepulveda, "Recycling Colonial Imaginaries: Havana's Restoration Project and the Nineteenth-Century Picturesque," in CAA 2020 Annual Conference (CAA, 2020).
15. Lapique Becali, *La Memoria en las Piedras,* 145–56.
16. In the Spanish version, the map is preceded by ten pages of small reproductions of historical maps, while in the French version, it is the first map in the atlas.
17. Rafael Rodríguez, *Atlas Cubano* (Havana: Litografía de la Real Sociedad Económica, 1841–1842).
18. Lapique Becali, *La Memoria en las Piedras,* 96, note 13.

19. Sophie Sapp Moore, Monique Allewaert, Pablo F. Gómez, and Gregg Mitman, "Plantation Legacies," January 22, 2019, *Edge Effects,* https://edgeeffects.net/plantation-legacies-plantationocene.
20. Jonathan Crary, "Géricault, the Panorama, and Sites of Reality in the Early Nineteenth Century," *Grey Room,* no. 9 (Autumn 2002): 5–25.
21. Zoila Lapique Becali, "Introduction," in Justo Germán Cantero, *Los Ingenios: Colección de Vistas de los Principales Ingenios de Azúcar de la Isla de Cuba,* Raros y Valiosas Colección Facsimilar (Havana: Biblioteca Nacional de Cuba José Martí, 2011).
22. J. G. Cantero, "Introduction," in *Los Ingenios: Colección de Vistas de los Principales Ingenios de Azúcar de la Isla de Cuba* (Havana: Impreso en la litografía de Luis Marquier, 1857), 1.
23. Cantero, "Introduction," 1.
24. Lapique Becali, *La Memoria en las Piedras,* 185.
25. Lapique Becali, "Introduction," i.
26. Lapique Becali, *La Memoria en las Piedras,* 184.
27. Luis Miguel García Mora and Antonio Santamaría García, "Donde Cristaliza la Esperanza: Lectura de los Ingenios," in *Los Ingenios: Colección de Vistas de los Principales Ingenios de Azúcar de la Isla de Cuba,* by Justo G. Cantero and Eduardo Laplante (Madrid: Centro Estudios y Experimentación de Obras Públicas, 2005), 64–65. For more on Cantero's ambition, see also Lorena Tezanos Toral, "The Architecture of Nineteenth-Century Cuban Sugar Mills: Creole Power and African Resistance in Late Colonial Cuba" (Ph.D. diss., City University of New York, 2015), chap. 2.
28. In this reading, I expand on that presented in John E. Crowley, "Sugar Machines: Picturing Industrialized Slavery," *American Historical Review* 121, no. 2 (April 2016): 403–36.
29. Antonio Benítez Rojo, *The Repeating Island: The Caribbean and the Postmodern Perspective,* trans. James E. Maraniss, 2nd ed., Post-Contemporary Interventions (Durham: Duke University Press, 1996), 113.
30. Gaceta de La Habana, July 11, 1855, quoted in Justo Germán Cantero, *Los Ingenios: Colección de Vistas de Los Principales Ingenios de Azúcar de la Isla de Cuba,* ed. Zoila Lapique Becali, Raros y Valiosas Colección Facsimilar (Havana: Biblioteca Nacional de Cuba José Martí, 2011), iv.
31. Cantero, "Introduction," 1.
32. Tim Barringer, Gillian Forrester, and Barbaro Martinez-Ruiz, eds., *Art and Emancipation in Jamaica: Isaac Mendes Belisario and His Worlds* (New Haven: Yale Center for British Art in association with Yale University Press, 2007), 54.
33. Barringer, Forrester, and Martinez-Ruiz, eds., *Art and Emancipation in Jamaica,* 347.
34. Tim Barringer, "Picturesque Prospects and the Labor of the Enslaved," in *Art and Emancipation in Jamaica,* ed. Barringer, Forrester, and Martinez-Ruiz, 45, and Geoff Quilley, "Pastoral Plantations: The Slave Trade and the Representation of British Colonial Landscape

in the Late Eighteenth Century," in *An Economy of Colour: Visual Culture and the Atlantic World, 1660–1830,* ed. Geoff Quilley and Kay Dian Kriz (Manchester: Manchester University Press, 2003), 106–28.

35. Esteban Montejo, *The Autobiography of a Runaway Slave,* ed. Miguel Barnet, trans. Jocasta Innes (London: Bodley Head, 1966), 108.

36. Charles Derosne and Jean-François Cail, "De la Elaboración del Azúcar en las Colonias y de los Nuevos Aparatos Destinados a Mejorala" (Madrid: Imprenta del Gobierno, 1844), 15, emphasis added, quoted in Daniel B. Rood, "A Creole Industrial Revolution in the Cuban Sugar Mill," in *The Reinvention of Atlantic Slavery: Technology, Labor, Race, and Capitalism in the Greater Caribbean* (Oxford: Oxford University Press, 2017), 10–11.

37. Rood, "A Creole Industrial Revolution," 18.

38. Cantero, "Introduction," 10.

39. Cantero, "Introduction," 10.

40. Cantero, "Introduction," 1.

41. Cantero, "Introduction," 1.

42. *Memorias,* vol. 2 (1836), especially issue 7, May 1836.

43. Evelyn Hu-DeHart, "Chinese Coolie Labor in Cuba in the Nineteenth Century: Free Labor of Neoslavery," *Contributions in Black Studies* 12: Ethnicity, Gender, Culture, Cuba (1994): 38–54.

44. Walton Look Lai, *Indentured Labor, Caribbean Sugar: Chinese and Indian Migrants to the British West Indies, 1838–1918* (Baltimore: Johns Hopkins University Press, 1993), 19. For a moving personal examination of this migration, see Gaiutra Bahadur, *Coolie Woman: The Odyssey of Indenture* (Chicago: University of Chicago Press, 2014).

45. *Anales de Ciencias, Agricultura, Comercio y Artes,* vol. 1 (Havana: Oficina del Gobierno y Capitanía General, 1827), 101; *Memorias de la Real Sociedad Patriótica de la Habana,* vol. 3 (1837), 110; *Anales de las Reales Junta de Fomento y Sociedad Económica de la Habana,* vol. 1 (Havana: Imprenta del Gobierno y de la Real Sociedad Económica, 1849), 245.

46. Cantero, "Introduction," 9.

47. Gillian McGillivray, *Blazing Cane: Sugar Communities, Class, and State Formation in Cuba, 1868–1959,* American Encounters/Global Interactions (Durham: Duke University Press, 2009), 21.

48. McGillivray, *Blazing Cane,* 14.

49. Mark Smith, "The Political Economy of Sugar Production and the Environment of Eastern Cuba, 1898–1923," *Environmental History Review* 19, no. 4 (1995): 31–48, https://doi.org/10.2307/3984691.

50. Quiroz, "Free Association and Civil Society in Cuba, 1787–1895."

51. Alison Fraunhar, *Mulata Nation: Visualizing Race and Gender in Cuba,* electronic resource, Caribbean Studies Series (Jackson: University Press of Mississippi, 2018).

5. What Do the Political and Ecological Layers Mean?

1. See especially Childs, *The 1812 Aponte Rebellion in Cuba;* Sartorius, *Ever Faithful;* Finch, *Rethinking Slave Rebellion in Cuba;* Márcia Regina Berbel, Rafael de Bivar Marquese, and Tâmis Parron, *Slavery and Politics: Brazil and Cuba, 1790–1850* (Albuquerque: University of New Mexico Press, 2016).
2. For more on these processions, see especially Anderson, *Carnival and National Identity in the Poetry of Afrocubanismo;* Yvonne Daniel, *Dancing Wisdom: Embodied Knowledge in Haitian Vodou, Cuban Yoruba, and Bahian Candomblé* (Urbana: University of Illinois Press, 2005); Walker, *No More, No More.* For more on how processions were used to perform complex group identity and reinforce power in the context of the kingdom of Kongo, see Cecille Fromont, *The Art of Conversion: Christian Visual Culture in the Kingdom of Kongo* (Chapel Hill, N.C.: The Omohundro Institute of Early American History and Culture by the University of North Carolina Press, 2014), chap. 1. See also Fernando Ortiz and Diana Iznaga, *Los Negros Curros* (Havana: Editorial de Ciencias Sociales, 1986); Lydia Cabrera, *El monte: Igbo Finda, Ewe Orisha, Vititinfinda* (Havana: Ed. C.R., 1954).
3. Fredrika Bremer, *The Homes of the New World: Impressions of America* (New York: Harper & Brothers, 1853); Mary Gardner Lowell, *New Year in Cuba: Mary Gardner Lowell's Travel Diary, 1831–1832,* ed. Karen Robert, The New England Women's Diaries Series (Boston: Massachusetts Historical Society and Northeastern University Press, 2003); Julia Ward Howe, *A Trip to Cuba* (Boston: Ticknor and Fields, 1860).
4. Adelaida de Juan, *Pintura Cubana: Temas y Variaciones* (Havana: UNEAC, 1978), 27.
5. Lowell, *New Year in Cuba,* introduction.
6. "'Jewel of Cuban Literature' Was Rediscovered, Documenting the Island in 19th Century," September 12, 2019, *WLRN: The South Florida Roundup,* https://www.wlrn.org/post/jewel-cuban-literature-was-rediscovered-documenting-island-19th-century, accessed 4/2/23.
7. Adriana Méndez Rodenas, *Gender and Nationalism in Colonial Cuba: The Travels of Santa Cruz y Montalvo, Condesa de Merlin* (Nashville: Vanderbilt University Press, 1998).
8. Verena Martinez-Alier and Verena Stolcke, *Marriage, Class, and Colour in Nineteenth-Century Cuba: A Study of Racial Attitudes and Sexual Values in a Slave Society* (Ann Arbor: University of Michigan Press, 1989); Gema R. Guevara, "Inexacting Whiteness: Blanqueamiento as a Gender-Specific Trope in the Nineteenth Century," *Cuban Studies* 36, no. 1 (December 28, 2005): 105–28; Karen Y. Morrison, "Slave Mothers and White Fathers: Defining Family and Status in Late Colonial Cuba," *Slavery & Abolition* 31, no. 1 (March 1, 2010): 29–55; Camillia Cowling, *Conceiving Freedom: Women of Color, Gender,*

and the Abolition of Slavery in Havana and Rio de Janeiro (Chapel Hill: University of North Carolina Press, 2013).

9. Michele Reid-Vazquez, "Tensions of Race, Gender, and Midwifery in Colonial Cuba," in *Africans to Spanish America: Expanding the Diaspora,* ed. Sherwin K. Bryant, Rachel Sarah O'Toole, and Ben Vinson (Urbana: University of Illinois Press, 2012), 186–205.
10. Olga Portuondo Zúñiga, *Entre Esclavos y Libres de Cuba Colonial* (Santiago de Cuba: Editorial Oriente, 2003); Lucero, *A Cuban City, Segregated.*

Index